Contemporary Holography

Multidisciplinary and Applied Optics

Series Editor
Vasudevan Lakshminarayanan
University of Waterloo, Ontario, Canada

Quantum Mechanics of Charged Particle Beam Optics: Understanding Devices from Electron Microscopes to Particle Accelerators
Ramaswamy Jagannathan, Sameen Ahmed Khan

Understanding Optics with Python
Vasudevan Lakshminarayanan, Hassen Ghalila, Ahmed Ammar, L. Srinivasa Varadharajan

Nonlinear Meta-Optics
Edited by Costantino De Angelis, Giuseppe Leo, Dragomir N. Neshev

Fourier Theory in Optics and Optical Information Processing
Toyohiko Yatagai

Contemporary Holography
C. S. Narayanamurthy

For more information about this series, please visit: https://www.routledge.com/Multidisciplinary-and-Applied-Optics/book-series/CRCMULAPPOPT

Contemporary Holography

C. S. Narayanamurthy

CRC Press is an imprint of the
Taylor & Francis Group, an **informa** business

First edition published 2023
by CRC Press
6000 Broken Sound Parkway NW, Suite 300, Boca Raton, FL 33487-2742

and by CRC Press
4 Park Square, Milton Park, Abingdon, Oxon, OX14 4RN

Library of Congress Cataloging-in-Publication Data

Names: Narayanamurthy, C. S., author.
Title: Contemporary holography / C.S. Narayanamurthy.
Description: Boca Raton : CRC Press, 2022. | Includes bibliographical references and index. | Summary: "This book is unique in covering most of developments on optical holography starting from photopolymer recording techniques to CMOS based digital holography. It also covers fundamentals and applications of dynamic holography using photorefractive crystals and many different types of digital holography and its many applications"-- Provided by publisher.
Identifiers: LCCN 2021043775 (print) | LCCN 2021043776 (ebook) | ISBN 9780367468279 (hardback) | ISBN 9781032181387 (paperback) | ISBN 9780367470975 (ebook)
Subjects: LCSH: Holography.
Classification: LCC TA1540 .N37 2022 (print) | LCC TA1540 (ebook) | DDC 621.36/75--dc23
LC record available at https://lccn.loc.gov/2021043775
LC ebook record available at https://lccn.loc.gov/2021043776

ISBN: 978-0-367-46827-9 (hbk)
ISBN: 978-1-032-18138-7 (pbk)
ISBN: 978-0-367-47097-5 (ebk)

DOI: 10.1201/9780367470975

Typeset in CMR10 font
by KnowledgeWorks Global Ltd.

Publisher's note: This book has been prepared from camera-ready copy provided by the authors.

Dedication

To my father and teachers

Contents

Preface

Writing a textbook on *holography* covering new techniques since its invention by Gabor in 1947 is a Himalayan task as the field has grown tremendously over the years. It will require several books to include all applications of holography as they have spread across several areas ranging from biology to astronomy in science and from civil engineering to aeronautical engineering in technology. In this book I try to elaborate various holographic techniques starting from conventional holography to the latest digital holography. One can easily distinguish the developments of holographic techniques in to two parts: one before the digital electronics revolution, and the second after that. Full fledged, conventional holography had to wait for 14 years after the initial discovery by Denis Gabor. This is because the first hologram could be recorded by Leith and Upatniks only after the invention of lasers in 1961. Since then, the holographic techniques and their applications have revolutionized many areas of science and technology. From 1961 to 1970 holographic recordings were carried out mainly using high resolution photographic plates which required wet processing for retrieving stored information in two and three dimensions. In the 1970s, due to the development of computing technologies especially, digital cameras for recording high quality images started replacing conventional wet processing of even ordinary photographs. Simultaneously a new field had emerged in optics, namely non-linear optics in the 1970's, and Professor Nicholus Bloombergen was awarded the Nobel Prize in 1971 for the same. At this juncture with intense laser beams and two wave mixing/four wave mixing in non-linear crystals had important applications like phase conjugation, Brillouin scattering, second harmonic generation, Raman scattering. Even though phase conjugation was responsible for the reconstructed real image of a hologram at that juncture it was considered as a real and virtual image of a recorded hologram. Only after phase conjugation was obtained in Barium Titanate($BaTiO_3$) using laser beam and then using two wave mixing did the phenomena became prominent. At that period another important development took place in non-linear optics, known as the photorefractive effect, in which two interfering beams can self diffract and create dynamic hologram of one of the beams. Photorefractive effect requires relative intensity of light in certain inorganic crystals like Bismuth Silicon Oxide($Bi_{12}SiO_{20}$), Bismuth Titanium Oxide($Bi_{12}TiO_{20}$) and Bismuth Germanium Oxide($Bi_{12}GeO_{20}$) or Polar crystals like Lithium Niobate($LiNBO_3$) and Barium Titanate($BaTiO_3$). Then another type of holography surfaced known as conoscopic holography and due to tremendous developments in CCD(Charge Coupled Devise)/CMOS cameras in the '90s the digital recording of holography emerged which completely revolutionised the holographic applications. In fact the industries which were hesitating to implement holographic technologies due to wet processing and the smaller sizes of holograms recorded using photorefractive crystals overwhelmingly accepted digital holography. Unfortunately, since the classic books on holography by Collier and Hariharan, no book is available and these two books were written in the '70s and '80s, respectively. Books by Collier and Hariharan mostly discussed conventional holography using photographic plates and holographic interferometric applications. Also, no coherent book on holography exists now which includes photorefractive holography, Conoscopic holography, Computer generated holography, and Digital holography. This book is an attempt to include all contemporary holographic techniques.

C.S. Narayanamurthy
Trivandrum, Kerala State, India
April 2020

Acknowledgments

I first want to acknowledge my late father, *Mr. C. S. Subramanian*, for encouraging me to write a textbook several years ago, when I first joined as faculty in physics. Unfortunately, I could not write it when he was alive. Then, I want to acknowledge all my teachers who taught me Applied Optics, especially my supervisor, *Prof. R. S. Sirohi*, who himself has written some excellent optics books and introduced holography to me in his applied optics laboratory at IIT Madras. *Prof. Sirohi* used to train us by asking each research scholar to give a talk on each chapter from Jenkins' and White's classical book on Optics, which further increased my interest in writing a book in one of the optics topics. Further, my sincere acknowledgment goes to *Prof. Chris Dainty*, for training me in the area of photorefractive holography in the Blacket Laboratory of Imperial College, London during 1990-1991. My post-doctoral days at Imperial, where *Prof. Denis Gabor* used to work and got his Nobel Prize, further increased my quest to write a book on holography. I also thank all my research and post-graduate students for constantly giving various inputs in my classes. I want to specially acknowledge the Department of Science and Technology (DST), Govt. of India for providing me BOYSCAST fellowship in 1990–91 and providing me with a project on dynamic holography during the 1995–1999 period and these two really propelled my interest in holography. I would also like to thank *Prof. Vasudevan Lakshmi Narayanan (Vengu)* of University of Waterloo, Canada for introducing me to *Marc Gutierrez* of CRC Press to write a book on holography. I acknowledge the patience of Marc in this regard for the past one year for accepting my demand for more time whenever I asked for it. Finally, I would like to thank *Prof. S. Murugesh* and *Dr. Sudheesh*, for helping me in solving my doubts in LaTeX and lastly, my student, *Dr. Pramod Panchal*, for the excellent drawing and giving me original experimental results which helped me to write this book during these difficult times. Finally, I would like to thank my wife, son and daughter for their continuous support in my all endeavors.

CSN

Contributors

C. S. Narayanamurthy
Dept. of Physics,
Indian Institute of Space Science and Technology(IIST), Valiamala (PO),
Trivandrum - 695547, INDIA

List of Figures

1 Conventional Holography

1.1 INTRODUCTION

Holography is perhaps the best experimental discovery since the invention of lasers in 1961. Though the original concept of holography was conceptulated in 1948 by British scientist Denis Gabor[1,2], real experimental demonstration had to wait till the invention of lasers by Charles Hard Townes in 1960. Previously we know that photography can record an image of an object which is basically two dimension. The real difference between understanding a 2 dimensional image and three dimensional image is to record information about phase of light beams. In common man' s language three dimension image is to record the depth information about our nose and ear using phase of the light beams. Normal photographs can not record the phase information as it records only the intensity of light beams reflected by any object without the phase information. Actually the light beams reflected by any object or person illuminated by it do contain phase information but the detecting film or CCD array records only the intensity and the reconstruction process will not be able to reconstruct the phase information from the intensity of light beam. At this juncture the use of laser beam due to its coherence(Appendix A) property play crucial role in holography for recording and reconstructing the phase information. The coherence property of light beam is essential for recording phase as light must fluctuate coherently in both space and time at the time of recording as well as reconstruction to recreate a time delayed image of an object. Though Gabor was interested in improving the quality of images in electron microscopy using the concept of reconstruction of a complete wavefront, he used the reference beam and tried to improve the quality of images in electron microscopy and that became crucial for recording 3D images in holography. Later after the invention of lasers simultaneously Leith and Upatnieks[3,5] and Denisyuk[4] developed real 3D of-axis holography. There are several books on holography[6-11] and most of them follow Fourier optics approach for theoretical descriptions and in this book we follow K K Sharma[11] style of Fourier optics explanation. In general, holographic recording can be of two types i) In-line or Gabor holography which is two dimensional and ii) Off-axis holography which is 3-Dimensional. Unlike photography, the process of holography requires two steps one for recording a hologram of an object in photographic plate and the other step for reconstructing the recorded object from the hologram. This process is common for both in-line as well as for off-axis holography.

1.2 IN-LINE GABOR HOLOGRAPHY

1.2.1 CONSTRUCTION OF HOLOGRAM

As described earlier holography has two stages construction and reconstruction of hologram. First we consider the construction of hologram of a point object and then

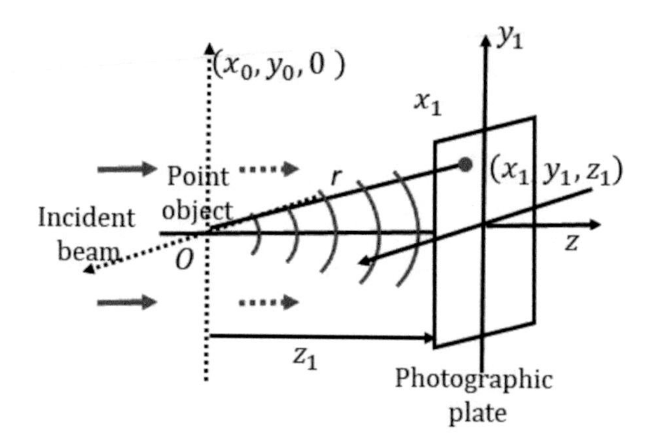

Figure 1.1 Experimental geometry for recording a point object Gabor hologram.

we extend the same to 2D transparent object. The experimental geometry necessary for recoding an in-line Gabor hologram of a point object is shown in Fig. 1.1, where, the incident collimated light beam from laser source illuminates the point object. Following Fig. 1.1, it shows a diffracted spherical beam emerges out from the point objecr and the other part of collimated beam shown is un-diffracted. The diffracted spherical beam from point object is the object beam and the un-diffracted beam acts as reference beam respectively. If $A(O) = \frac{A_O}{r} e^{ik_1 r}$ represents the scattered spherical object beam from the point source and $A(R) = A_R e^{ik_1 z_1}$ represents un-diffracted reference beam then the total complex light field at the photographic plate will be,

$$A(x_1, y_1, z_1) = \frac{A_O}{r} e^{ik_1 r} + A_R e^{ik_1 z_1} \tag{1.1}$$

This is the first step in recording hologram. The interference pattern of object and reference beam shown in Eqn.1.1 is recorded in the photographic plate shown in Fig. 1.1 and while doing so no temporary changes or spatial disturbances are allowed as it will destroy the recording of phase of the object beam from the point object. For that vibration isolation table is needed but if the laser beam is pulsed and with its duration longer than temporal fluctuation of object then, vibration isolation table is not required. Now the total intensity recorded by the photographic plate kept at a distance z_1 can be written as,

$$I(x_1, y_1, z_1) = \left|\frac{A_O}{r} e^{ik_1 \left(z_1 + \frac{x_1^2 + y_1^2}{2z_1}\right)} + A_R e^{ik_1 z_1}\right|^2 \tag{1.2}$$

where $r = z_1 + \frac{x_1^2 + y_1^2}{2z_1}$ and Eqn.1.2 becomes,

$$I(x_1, y_1, z_1) = \left(\frac{1}{2}\varepsilon_0 c\right) \left[\frac{|A_O|^2}{z_1^2} + |A_R|^2 + \frac{A_O A_R^*}{z_1} e^{\frac{ik_1}{2z_1}(x_1^2 + y_1^2)} + \frac{A_O^* A_R}{z_1} e^{\frac{-ik_1}{2z_1}(x_1^2 + y_1^2)}\right] \tag{1.3}$$

$$= \left(\frac{1}{2}\varepsilon_0 c\right)\left[\frac{|A_O|^2}{z_1^2} + |A_R|^2 + 2\frac{|A_O||A_R|}{z_1} cos\left(\frac{k_1}{2z_1}(x_1^2+y_1^2)+\phi_0\right)\right] \quad (1.4)$$

Equation 1.4 thus represents total intensity recorded and the term ϕ_0 is the phase difference between object and reference beams. The photographic plate records all the spherical waves emanating from point source along with the reference beam thus creating interference pattern. Depending upon the power of laser beam the exposure time varies for recording the hologram. Normally the reference beam intensity may be higher than object beam as object beam is diffracted spherical beam where as reference beam is un-diffracted collimated beam. The film exposure is equal to $I\Delta t$ with Δt giving the exposure time. The hologram recording photographic film/plate needs to be highly sensitive to the recording light wavelength and must have high resolving capability of rapid spatial intensity variations of the intensity pattern. There are high sensitive holographic recording plates for example the Agfa Gavert's 10E 75/8E 75 for recording holograms and volume holograms. After recording the holograms, the photographic film/plate is developed/fixed and bleached. The processed holographic plate/film will have an amplitude transmittance equal to,

$$T(x_1,y_1,z_1) = T_0 - \alpha I\Delta t = A_0 + B_0 e^{\frac{ik_1}{2z_1}(x_1^2+y_1^2)} + B_0^* e^{\frac{-ik_1}{2z_1}(x_1^2+y_1^2)} \quad (1.5)$$

$$= A_0 + 2|B_0| cos[\frac{k_1}{2z_1}(x_1^2+y_1^2)+\phi_0] \quad (1.6)$$

where, $A_0 = A + B(\frac{|A_O|^2}{z_1^2} + |A_R|^2)$ and $B_0 = B\frac{A_O A_R^*}{z_1}$. Eqn.1.5 represents hologram of the point object where, the first term will attenuate the beam passing through it similar to semi transparent glass plate. The second and third terms are actually represent the real and virtual images of the point object and they are similar to concave and convex lenses respectively. Thus hologram of many such point objects along the optical axis can generate Fresnel lenses of positive and negative lensing actions.

1.2.2 RECONSTRUCTION OF HOLOGRAM USING NORMAL INCIDENCE OF REFERENCE BEAM

The reconstruction of processed hologram can be by normal incidence of reference beam. The relevant experimental geometry for reconstruction of processed hologram is shown in Fig. 1.2. The reference beam will reconstruct both virtual and real images of the point object. The field distribution at a plane z_2 behind the processed hologram(Fig 1.2) is given by,

$$A_H(x_2,y_2,z_1+z_2) = \left[\frac{-i}{\lambda_1}A_h e^{\frac{ik_1(z_1+z_2)}{z_2}} T(x_2,y_2,z_1) * h_{z_2}(x_2,y_2)\right] \quad (1.7)$$

$$= C\int\int_{-\infty}^{+\infty} T(x_1,y_1,z_1) e^{\frac{ik_1}{2z_2}(x_2-x_1)^2+(y_2-y_1)^2} dx_1 dy_1 \quad (1.8)$$

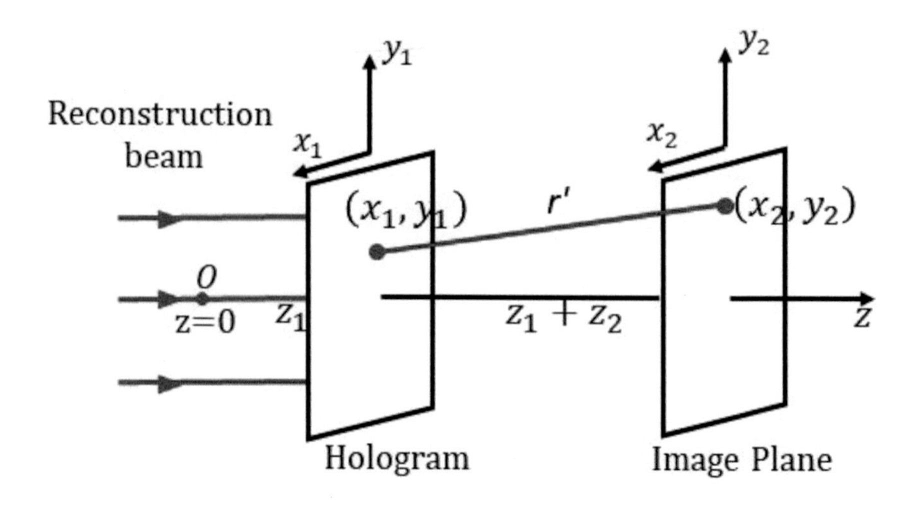

Figure 1.2 Wavefront reconstruction of scattered light beam from point source using plane reference beam.

Expanding further Eqn.1.8 becomes,

$$= [CA_h \int\int_{-\infty}^{\infty} e^{\frac{ik_1}{2z_2}(x_2-x_1)^2+(y_2-y_1)^2} dx_1 dy_1$$

$$+CB_0 e^{\frac{ik_1}{2z_2}(x_2^2+y_2^2)}$$

$$\int\int_{-\infty}^{+\infty} e^{\frac{i}{2}(\frac{k_1}{z_1}+\frac{k_1}{z_2})(x_1^2+y_1^2)} e^{\frac{-ik_1}{z_2}(x_1x_2+y_1y_2)} dx_1 dy_1$$

$$+CB_0^* e^{\frac{ik_1}{2z_2}(x_2^2+y_2^2)}$$

$$\int\int_{-\infty}^{+\infty} e^{\frac{i}{2}(\frac{-k_1}{z_1}+\frac{k_1}{z_2})(x_1^2+y_1^2)} e^{\frac{-ik_1}{z_2}(x_1x_2+y_1y_2)} dx_1 dy_1] \quad (1.9)$$

In Eqn.1.9 the value of term C is equal to $\frac{-i}{\lambda_1}\frac{A_0}{z_2}e^{ik_1(z_1+z_2)}$. Eqn.1.9 gives the reconstructed wavefront from the processed hologram where the first term is attenuated direct beam and the second and third terms give the object and its conjugate representing real and virtual images respectively. Substituting $z_1 = z_2$ i.e if the reconstructing plane or image plane is kept at z_1 behind the processed hologram which is same distance between point object and holographic plate while recording, then Eqn.1.9 will be equal to,

$$A_H(x_2, y_2, 2z_1) = C_1 + CB_0 e^{\frac{ik_1}{2z_1}(x_2^2+y_2^2)}$$

$$\int\int_{-\infty}^{\infty} e^{\frac{ik_1}{z_1}(x_1^2+y_1^2-x_1x_2-y_1y_2)} dx_1 dy_1$$

$$+CB_0' e^{\frac{ik_1}{2z_1}(x_2^2+y_2^2)} \int\int_{-\infty}^{\infty} e^{\frac{-ik_1}{z_1}(x_1^2+y_1^2-x_1x_2-y_1y_2)} dx_1 dy_1 \quad (1.10)$$

$$= C_1 + C_2 e^{\frac{ik_1}{4z_1}(x_2^2+y_2^2)} + C_3\delta(x_2, y_2) \tag{1.11}$$

Thus Eqn.1.11 represents the reconstructed wavefronts of processed hologram at the reconstruction or image plane. The 3rd term represents the real image of point represented by delta function and second term the virtual image of the point object which is shown in Fig. 1.3 and C_1, C_2, C_3 represent complex constants. In fact the

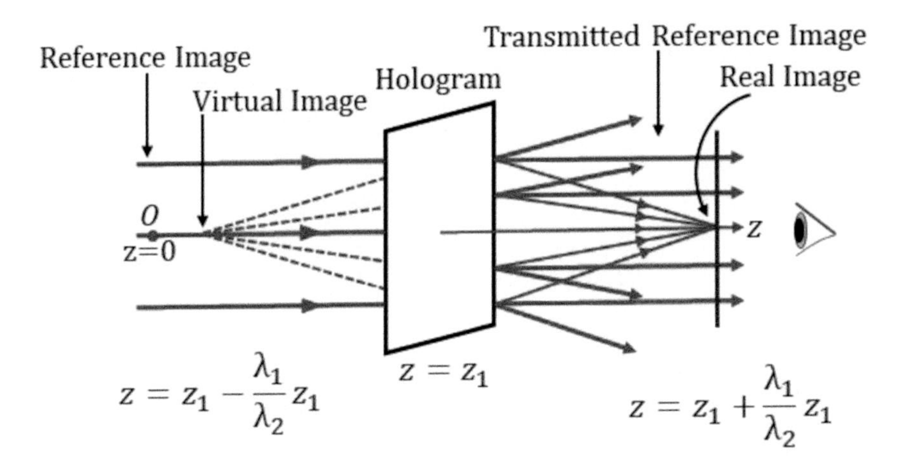

Figure 1.3 Reconstruction of real and virtual images from a recorded Gabor (inline) hologram of a point source.

real image is formed at a distance $z = 2z_1$ behind the processed hologram and virtual image at the same distance behind the hologram(Fig. 1.3). In real sense it is difficult to view real and virtual images as they form along same on-axis and all three i.e, attenuated reference beam represented by first term in Eqn.1.10, the real and virtual images can not be separated and in grating language the zeroth order(Attenuated wavefront), $+1$ order(real image) and -1 order(virtual image) all lie along on-axis. This can be corrected using Off-axis geometry with oblique incidence of reference beam.

1.3 OFF-AXIS HOLOGRAPHY

1.3.1 CONSTRUCTION OF OFF-AXIS HOLOGRAPHY

The main problem in in-line Gabor holography is separation of real and virtual images from the direct beam while reconstruction. This can be solved if an off-axis geometry is considered with object and reference beams, not in-line but off-axis with respect to each other. Fig 1.4 shows a typical off-axis holography geometry employed first by Leith and Upatnieks where the incident laser beam is split in to two by beam splitter. One beam from the beam splitter illuminates the point object and the other beam which is reference, illuminates the holographic plate at an off-axis

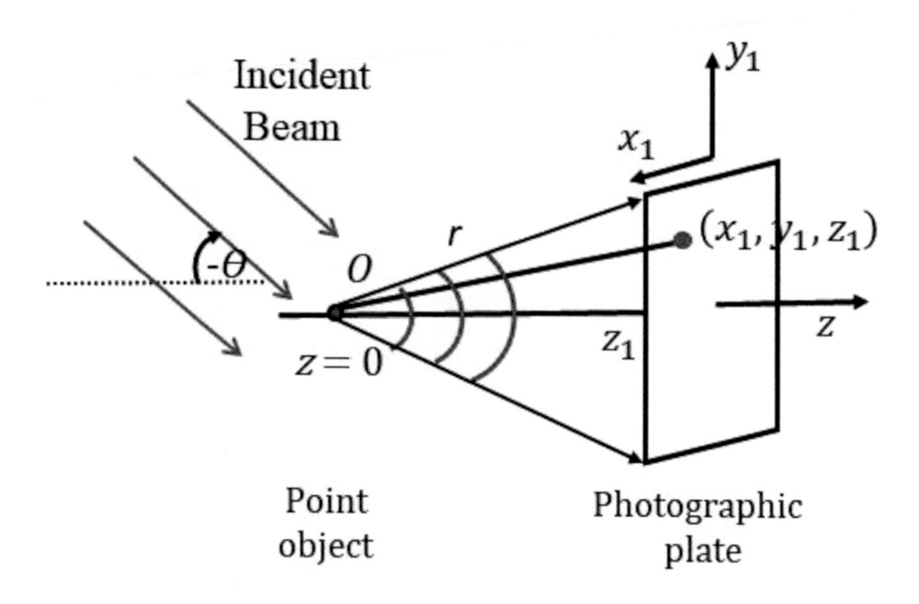

Figure 1.4 Recording of point object in an off-axis holography.

angle$-\theta$ as shown in Fig 1.4. The total field distribution at the plane of holographic plate is,

$$A_H(x_1,y_1,z_1) = \frac{A_O}{z_1} e^{ik(z_1+\frac{x_1^2+y_1^2}{2z_1})} + A_R e^{ik(z_1 cos\theta - x_1 sin\theta)} \tag{1.12}$$

The corresponding intensity distribution will be,

$$I_H(x_1,y_1,z_1) = [< (\frac{1}{2}\varepsilon_0 c) >) \frac{|A_O|^2}{z_1} + A_R^2 + \frac{A_O A_R^*}{z_1} e^{ikz_1(1-cos\theta)} \\ e^{ik(\frac{x_1^2+y_1^2}{2z_1}+x_1 sin\theta)} + \frac{A_O^* A_R}{z_1} e^{-ikz_1(1-cos\theta)} e^{-ik(\frac{x_1^2+y_1^2}{2z_1}+x_1 sin\theta)}] \tag{1.13}$$

Where A_H represents the amplitude of spherical wave emanating from the point object and A_R represents the amplitude of reference beam. The intensity of interference between object and reference beam represented by the Eqn. 1.13 is recorded by the holographic plate kept at a distance z_1 behind the point object is then processed(After developing and fixing). The processed holographic plate will now have the amplitude transmittance equal to,

$$T(x_1,y_1,z_1) = A_D + B_o e^{ik(\frac{x_1^2+y_1^2}{2z_1}+x_1 sin\theta)} + B_o^* e^{-ik(\frac{x_1^2+y_1^2}{2z_1}+x_1 sin\theta)} \tag{1.14}$$

where $B_o = \frac{A_H A_R^*}{z_1} e^{ikz_1(1-cos\theta)}$.

1.3.2 RECONSTRUCTION OF OFF-AXIS HOLOGRAPHY

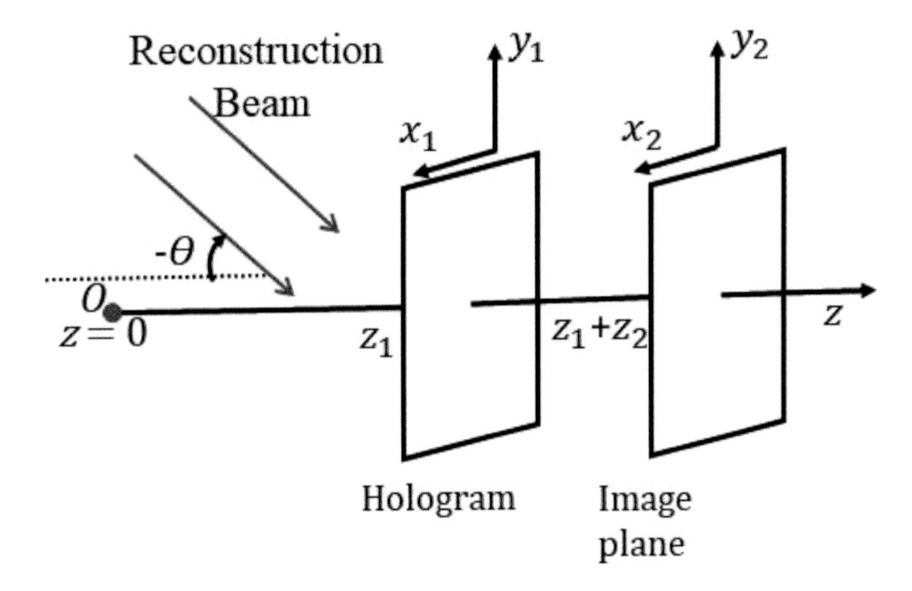

Figure 1.5 Reconstruction of the off-axis holographic point image using an oblique reference beam.

The reconstruction geometry for off-axis holography is shown in Fig. 1.5 where the same reference beam is used. The amplitude distribution of processed hologram in a plane at a distance z_2 behind it is given by,

$$E_H(x_1,y_1,z_1) = E_h e^{ik(z_1 cos\theta - x_1 sin\theta)} T(x_1,y_1) \tag{1.15}$$

$$E_H(x_2,y_2,z_1+z_2) = \frac{-i}{\lambda}\frac{e^{ikz_2}}{z_2} E_H(x_1,y_1,z_1) * h_{z_2}(x_2,y_2)$$
$$= \frac{-i}{\lambda}\frac{e^{ikz_2}}{z_2} \int\int_{-\infty}^{+\infty} E_H(x_1,y_1,z_1 e^{\frac{ik}{z_2}[(x_2-x_1)^2+(y_2-y_1)^2]} dx_1 dy_1 \tag{1.16}$$

Substituting Eqns 1.14 and 1.15 in Eqn 1.16 and simplifying we get,

$$E_H(x_2,y_2,z_1+z_2) = CA_d \int\int_{-\infty}^{+\infty} e^{\frac{ik}{2z_2}[(x_2-x_1)^2)-2z_2 x_1 sin\theta+(y_2-y_1)^2]} dx_1 dy_1$$
$$+CB_o \int\int_{-\infty}^{+\infty} e^{\frac{ik}{2z_1}[(x_1^2+y_1^2)+\frac{z_1}{z_2}((x_2-x_1)^2+(y_2-y_1)^2)]} dx_1 dy_1$$
$$+CB_o^* \int_{-\infty}^{+\infty} e^{\frac{-ik}{2z_1}[x_1^2+4z_1 x_1 sin\theta-\frac{z_1}{z_2}(x_2-x_1)^2]} dx_1$$
$$\int_{-\infty}^{+\infty} e^{\frac{-ik}{2z_1}[y_1^2-\frac{z_1}{z_2}(y_2-y_1)^2]} dy_1 \tag{1.17}$$

In Eqn 1.17, the constant term $C = \frac{-i}{\lambda} E_h \frac{e^{ik(z_1 cos\theta + z_2)}}{z_2}$. Simplifying further the field distribution behind the hologram plane which is at a distance z_1 can be written as,

$$E_H(x_2, y_2, 2z_1) = C_1 e^{-ikx_2 sin\theta} + C_2 e^{\frac{ik}{4z_1}(x_2^2 + y_2^2)} + C_3 \delta(x_2 + 2z_1 sin\theta, y_2) \quad (1.18)$$

Where C_1, C_2 and C_3 are complex constants. The first term in Eqn.1.18 represents the un-diffracted direct beam which is attenuated and propagates in same direction as reconstruction beam. The second and third terms represent virtual and real images of original object in front and behind the hologram as shown in Fig. 1.6. It can be

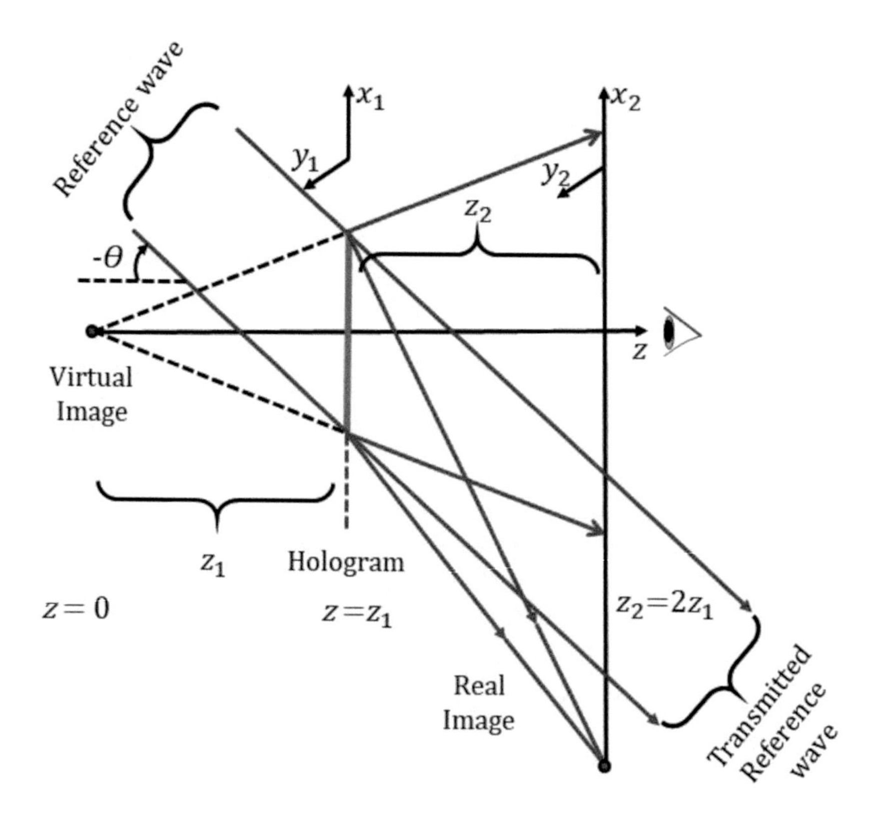

Figure 1.6 Reconstruction of the off-axis holographic point image using reconstruction beam at same inclination as reference beam.

seen that the real image can be clearly separated from background of virtual image by choosing proper choice of inclination angle of reference beam. Fig 1.6 and Fig 1.7 show reconstructed and separated real and virtual images of hologram of a point source, where the reference beam while recording was kept at an angle $-\theta$ and reconstruction wave at an angle $+\theta$ respectively.

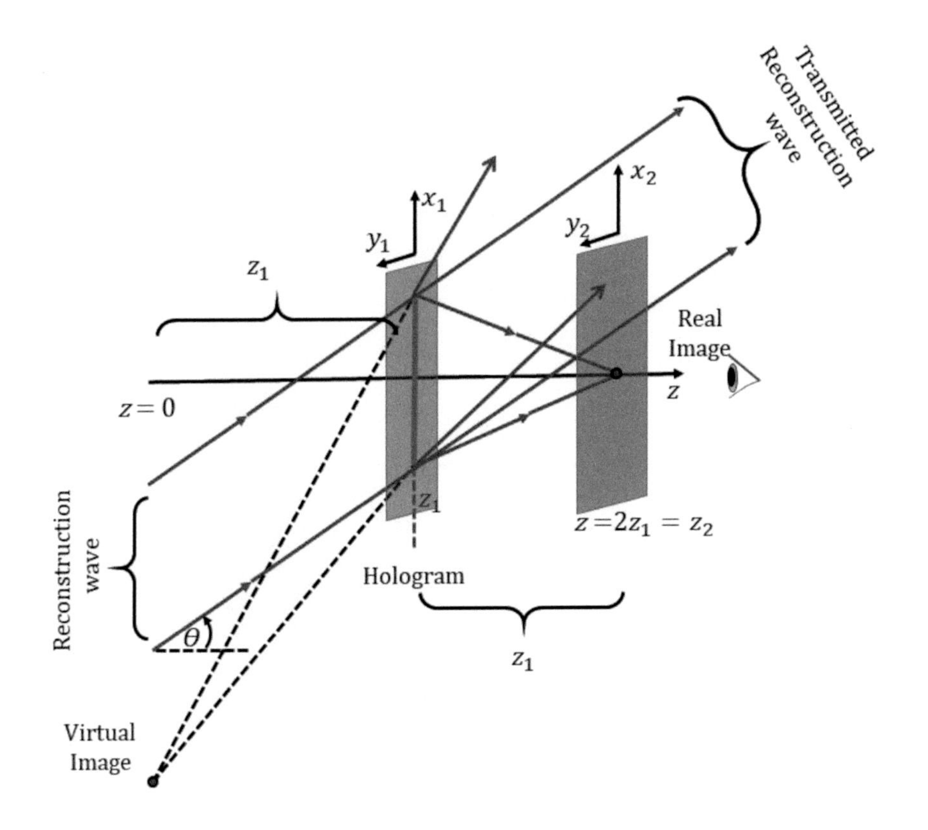

Figure 1.7 Reconstruction of the off-axis holographic virtual and real point images using a reconstruction beam at θ when reference beam was at $-\theta$ while recording.

1.4 POLARIZATION BASED HOLOGRAPHY

For recording holograms polarization property of light is also a necessary condition. This is because in constructing a hologram first we record the interference of object and reference beams on a photographic plate. To record good quality interference pattern of two beams, we require three basic conditions to be satisfied namely i) Both interfering beams have to be coherent spatially and temporally ii) Have to be same state of polarization iii) Should have same wavelengths (derived from same source). Recently there have been several research papers on unification of coherence and polarization of light beams especially from Emil Wolf and his group and one of closest experimental proof could be holography as these two properties of light are the first requirement for recording holograms both in photographic medium as well as in digital medium. In previous section we considered recording and reconstruction of holograms by not considering polarization states of light beam and in this section we introduce polarization state as they are important especially for recording phase and photoelastic (Stress birefringent) objects.

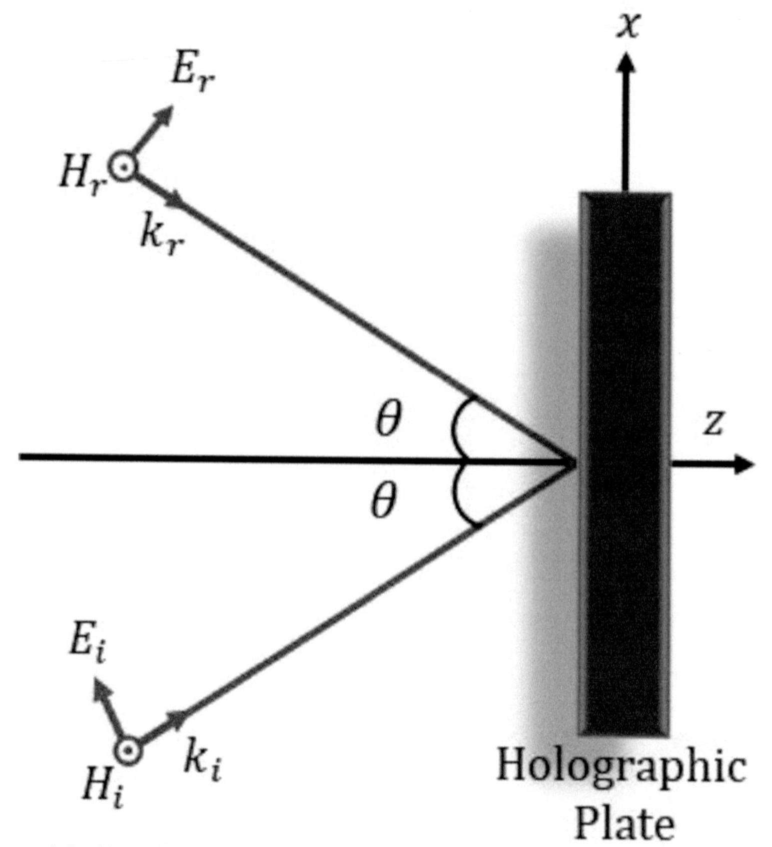

Figure 1.8 Recording geometry for incident object reference beams with their electric field components with parallel polarization to the incident plane.

1.4.1 PARALLEL POLARIZATION

Consider a linearly polarized light field from phase object with its electric field vector parallel to plane of incidence of a photographic plate and a reference beam with its electric field vector parallel to plane of incidence of same photographic plate interfere as shown in Fig 1.8. The light field amplitudes of electric field components of object beams(Fig 1.8) is,

$$\vec{E_O} = E_o(\hat{x}cos\theta - \hat{z}sin\theta)e^{-i\beta_1(xsin\theta+zcos\theta)}e^{i\phi_o} \tag{1.19}$$

The magnetic field component of incident object beam interms of electric field component can be written (Fig.) as,

$$\vec{H_O} = \frac{E_o}{\eta_o}e^{-i\beta_1(xsin\theta+zcos\theta)}\hat{y}e^{i\phi_o} \tag{1.20}$$

Then for the reference beam the electric field component is given by,

$$\vec{E_R} = E_r(\hat{x}cos\theta + \hat{z}sin\theta)e^{-i\beta_1(xsin\theta+zcos\theta)}e^{i\phi_r} \tag{1.21}$$

similarly for magnetic field component of reference beam we get,

$$\vec{H_R} = \frac{E_R}{\eta_o} e^{-i\beta_1)(xsin\theta + zcos\theta)} \hat{y} e^{i\phi_r} \tag{1.22}$$

In all electric and magnetic field components of object and reference beams, the first exponential term represents propagation vector and the terms $e^{i\phi_o}$ and $e^{i\phi_r}$ represent phase of object and reference beams respectively. Unlike conventional interference the intensity pattern of these two beams on the photographic plate can be written using Poynting's theorem in a new approach,

$$I_H = \vec{E_O} \times \vec{H_O^*} + \vec{E_R} \times \vec{H_R^*} + \vec{E_O}.\vec{E_R^*} + \vec{E_O^*}.\vec{E_R} \tag{1.23}$$

Substituting respective values of electric and magnetic field components we get following expression,

$$I_H = [\frac{1}{2}\varepsilon_0 c[(E_O^2(cos\theta\hat{z} + sin\theta\hat{x}) + E_R^2(cos\theta\hat{z} - sin\theta\hat{x})]$$
$$+ E_O E_R^*(cos^2\theta - sin^2\theta)e^{+i(\phi_0 - \phi_r)} + E_O^* E_R(cos^2\theta - sin^2\theta)e^{-i(\phi_0 - \phi_r)} \tag{1.24}$$

The hologram is reconstructed with the reference beam $\vec{E_R}$ given by Eqn.1.21 and the reconstructed intensity of hologram will become equal to,

$$I_{H_o} = [\frac{1}{2}\varepsilon_0 c[(2E_O^2 sin2\theta] + +4E_o E_r cos2\theta cos(\phi_o - \phi_r)].e^{-i\beta_1(xsin\theta + zcos\theta)} \tag{1.25}$$

The above Eqn.1.25 shows that the hologram can be reconstructed by changing the oblique incidence angle value θ. The reconstruction also shows that if, the electric fileds of incident object and reference beams areparallel to the plane of incidence on photographic plate then while reconstruction the reference beam intensity vanishes leaving only object beam intensity which can be avoided by choosing proper incidence angle.

1.4.2 PERPENDICULAR POLARIZATION

Consider a linearly polarized light field from phase object with its electric field vector perpendicular to plane of incidence of a photographic plate and a reference beam with its electric field vector perpendicular to plane of incidence of same photographic plate interfere as shown in Fig 1.9. The electric and magnetic field components of object and reference beams are given by,

$$\vec{E_O} = E_o e^{-i\beta_1(xsin\theta + zcos\theta)} \hat{y} e^{i\phi_o} \tag{1.26}$$

and its magnetic field component is equal to,

$$\vec{H_O} = \frac{E_o}{\eta_0}(-\hat{x}cos\theta + \hat{z}sin\theta)e^{-i\beta_1(xsin\theta + zcos\theta)} e^{i\phi_o} \tag{1.27}$$

Similarly for reference beam the electric field component can be written as,

$$\vec{E_R} = E_r e^{-i\beta_1(xsin\theta + zcos\theta)} \hat{y} e^{i\phi_r} \tag{1.28}$$

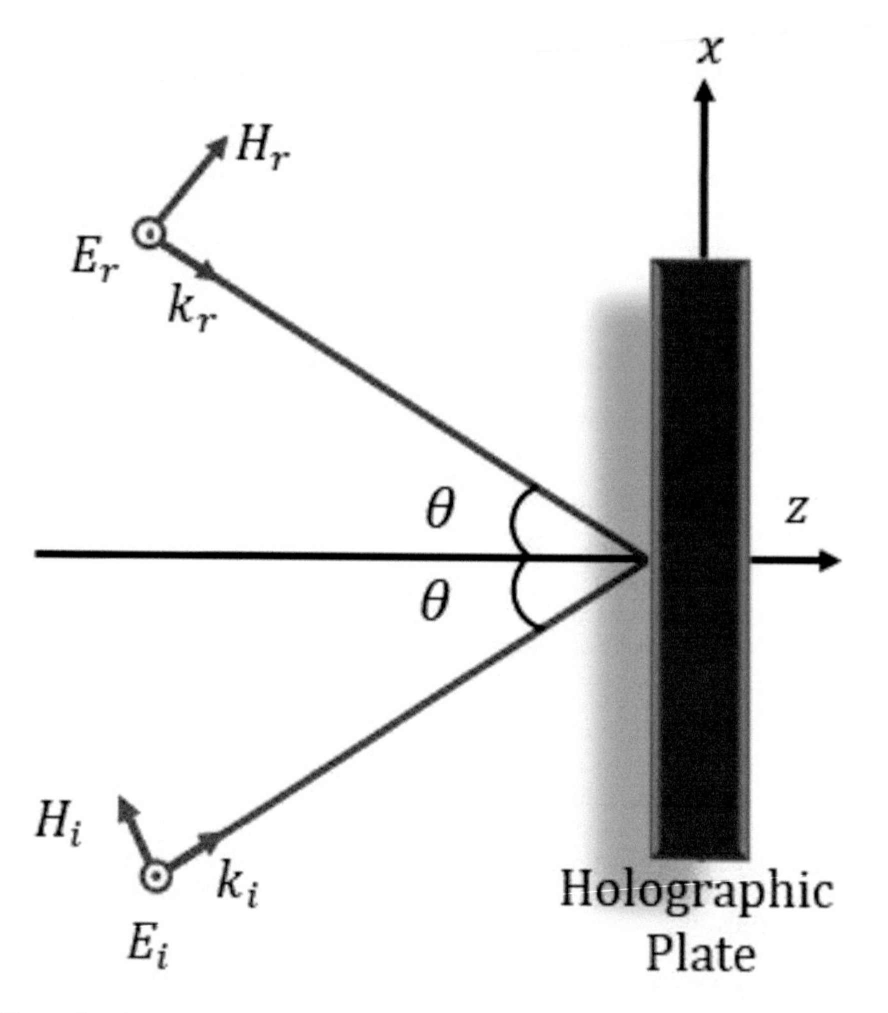

Figure 1.9 Recording geometry for incident object reference beams with their electric field components having perpendicular polarization to the incident plane.

and its magnetic field component equal to,

$$\vec{H}_R = \frac{E_o}{\eta_0}(\hat{x}cos\theta + \hat{z}sin\theta)e^{-i\beta_1(xsin\theta + zcos\theta)}e^{i\phi_r} \tag{1.29}$$

Now, the total intensity on the photographic plate can be written using Poynting's theorem for self intereference we get,

$$I_H = \frac{1}{2}\varepsilon_0 c[(E_O^2(cos\theta\hat{z} + sin\theta\hat{x}) + E_r^2(-cos\theta\hat{z} + sin\theta\hat{x})] + E_o E_r^* e^{i(\phi_o - \phi_r)} + E_o^* E_r e^{-i(\phi_o - \phi_r)} \tag{1.30}$$

Reconstructing Eqn.1.30 with the reference beam $\vec{E}_R$ we finally get,

$$I_{H_o} = E_r^2(E_0 e^{i\phi_o} + E_o^* e^{-i\phi_o}) \quad (1.31)$$

The final holographic equation 1.31 shows very clearly that if, the incident object and reference beam are having perpendicular polarization with respect to incident plane,then one can get clear object fields without need of any suppression of direct beams as they get cancelled.

1.5 OFF-AXIS HOLOGRAPHY WITH 3D OBJECTS

Though Denis Gabor invented the concept of holography in 1947, it was Leith and Upatnieks[3,5] who actually demonstrated holography with 3D objects. Consider Fig 1.10, where the expanded laser beam is split in to two by a beamsplitter (50 : 50) and one of beams falls on a 3D object. Now, consider a point O with position co-ordinates (x, y, z) on the surface of the object. The reflected laser beam from the object surface falls on a recording holographic plate which, is kept at a distance z_1 from the object. Simultaneously, an oblique reference beam of amplitude A_r making an angle θ with respect to axis of holographic plate also falls on the holographic plate. Thus both object beam and reference beam is allowed to interfere at the holographic plate. Care must be taken to see that the intensities of both object and reference beams are equal for efficient recording of hologram. Now, the total field distribution of both object and reference beam is given by,

$$A_H = A_o(x_1, y_1, z_1) + A_r e^{ik(z_1 cos\theta + x_1 sin\theta)} \quad (1.32)$$

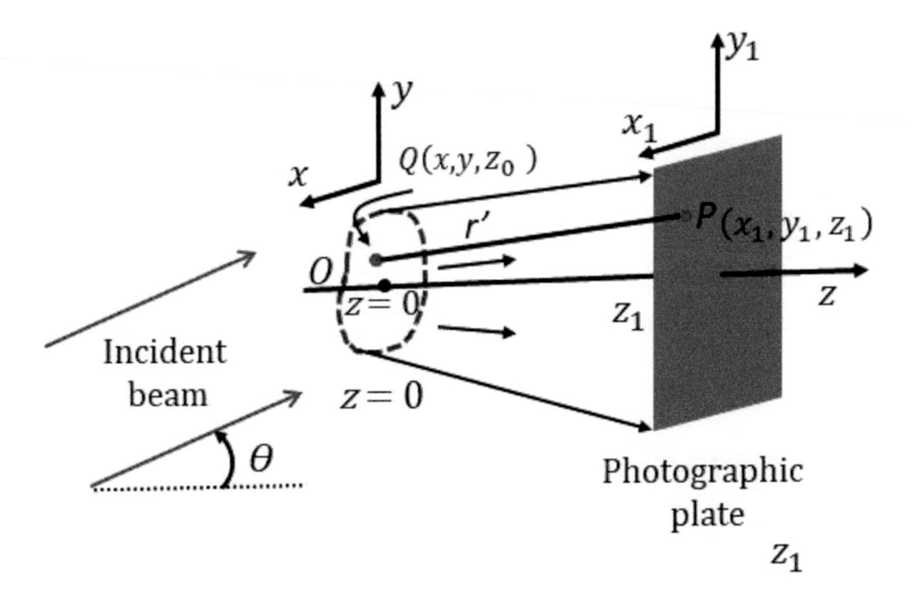

Figure 1.10 Recording of a 3D object in an off-axis geometry.

Using Fourier optics the object beam $A_0(x_1, y_1, z_1)$ is nothing but convolution of field distribution $A_o(x, y, z)$ on the surface of object with the impulse response function of space between the object and recording holographic plate and that can be written as,

$$A_0(x_1, y_1, z_1) = \frac{-i}{\lambda} \int \int_S A_0(x, y, z) e^{\frac{ik}{2(z_1 - z)}[(x_1 - x)^2 - (y_1 - y)^2]} dS \tag{1.33}$$

Then the total intensity distribution on the holographic plate is,

$$I(x_1, y_1, z_1) = |A_0(x_1, y_1, z_1)|^2 + |A_r|^2 + A_0(x_1, y_1, z_1) A_r^* e^{-ik(z_1 cos\theta + x_1 sin\theta)} + A_0^*(x_1, y_1, z_1) A_r^{ik(z_1 cos\theta + x_1 sin\theta)} \tag{1.34}$$

The recorded holographic plate is developed and fixed similar to developing conventional photographic plate and then is kept at same position for reconstruction. The amplitude transmittance of recorded holographic plate is given by,

$$A_T(x_1, y_1, z_1) = A_1 + B_1 |A_0(x_1, y_1, z_1)|^2 + B_2 e^{-ikx_1 sin\theta} A_0(x_1, y_1, z_1) + B_2^* e^{ikx_1 sin\theta} A_o^*(x_1, y_1, z_1) \tag{1.35}$$

In above equation the terms A_1, B_1, B_2 are constants. Now, when this developed, fixed and bleached hologram is illuminated using same reference beam as shown in Fig. 1.11, the field distribution at a distance z_2 behind the hologram is given by the following expression,

$$A_H(x_1, y_1, z_1) = A_r e^{ik(z_1 cos\theta + x_1 sin\theta)} A_T(x_1, y_1, z_1) \tag{1.36}$$

and then at the reconstruction plane (x_2, y_2, z_2) the field distribution is given by,

$$A_H(x_2, y_2, z_2) = \frac{-i}{\lambda} \frac{e^{ikz_2}}{z_2} A_r e^{ik(z_1 cos\theta + x_1 sin\theta)} A_T(x_2, y_2, z_1) * h_{z_2}(x_2, y_2) \tag{1.37}$$

$$\begin{aligned} &= A_1' \int \int_{-\infty}^{+\infty} e^{\frac{ik}{2z_2}[(x_2 - x_1)^2 + (y_2 - y_1)^2 + 2z_2 x_1 sin\theta]} dx_1 dy_1 \\ &+ B_1' \int \int_{-\infty}^{+\infty} |A_o(x_1, y_1, z_1)|^2 e^{\frac{ik}{2z_2}[(x_2 - x_1)^2 + (y_2 - y_1)^2 + 2z_2 x_1 sin\theta]} dx_1 dy_1 \\ &+ B_2' \int \int_{-\infty}^{+\infty} A_o(x_1, y_1, z_1) e^{\frac{ik}{2z_2}[(x_2 - x_1)^2 + (y_2 - y_1)^2]} dx_1 dy_1 \\ &+ B_2'^* \int \int_{-\infty}^{+\infty} A_o^*(x_1, y_1, z_1) e^{\frac{ik}{2z_2}[(x_2 - x_1)^2 + (y_2 - y_1)^2 + 4z_2 x_1 sin\theta]} dx_1 dy_1 \end{aligned} \tag{1.38}$$

In Eqn. 1.38, the $A_1', B_1', B_2', B_2'^*$ represent complex constants. The first term in Eqn. 1.38 represents transmitted reconstruction wave and second term spread out zero order diffracted beam respectively.

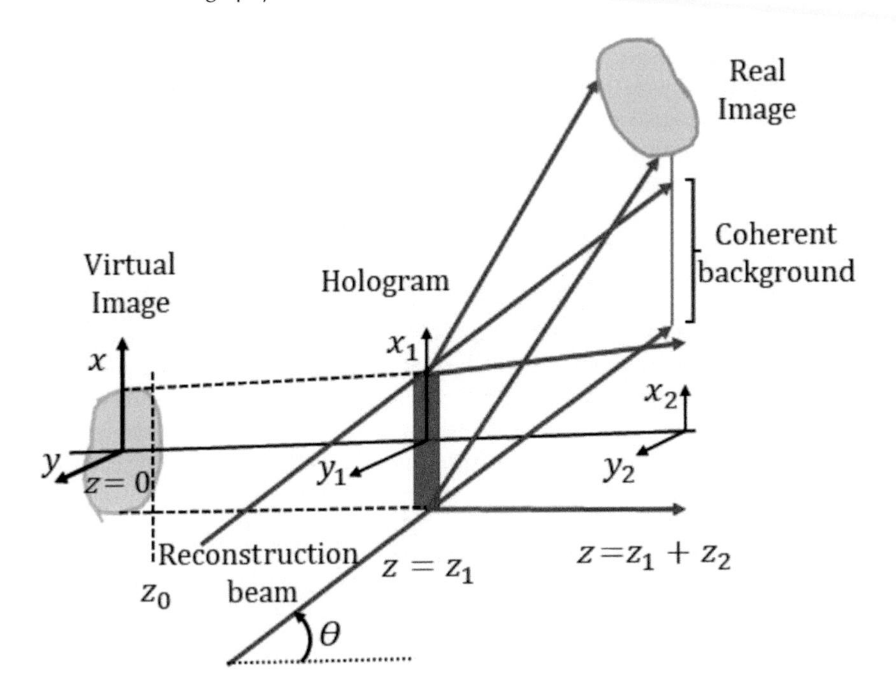

Figure 1.11 Reconstruction of the off-axis holographic virtual and real 3D object holographic images using a reconstruction beam at an angle θ.

1.5.1 RECONSTRUCTION OF VIRTUAL AND REAL IMAGES

To reconstruct the real image from the hologram consider the third term of Eqn.1.38 and substitute Eqn.1.33 in to it and that gives,

$$A_V(x_2,y_2,z_2) = \frac{-i}{\lambda} e^{\frac{ikz_1}{z_1}} B_2' e^{\frac{ik}{2z_2}(x_2^2+y_2^2)} \int\int_{-\infty}^{+\infty} A_o(x,y,z) e^{\frac{ik}{2(z_1-z)}(x^2+y^2)} \int\int_{-\infty}^{+\infty} e^{\frac{ik}{2}(\frac{1}{z_1-z}+\frac{1}{z_2})(x_1^2+y_1^2)} e^{-ik[x_1(\frac{x}{(z_1-z)}+\frac{x_2}{z_2})+y_1(\frac{y}{(z_1-z)}+\frac{y_2}{z_2})]} dx_1 dy_1 dx dy \quad (1.39)$$

The equation 1.39 gives the object field distribution of virtual image. It can be seen from Eqn. 1.39 the integration over the surface of object is replaced with integration over the transverse planes of the object. If we set $z_2 = -(z_1 - z)$ in the plane $z_1 = z$ then Eqn.1.30 reduces to,

$$A_V(x_2,y_2,z_1) = C_1 A_O(x_2 = x_1, y_2 = y_1, z_1 = z) \quad (1.40)$$

The constant C_1 is complex constant and the virtual image reproduces original field distribution of the object. The 4th term in Eqn.1.38 represents the real image which is formed due to phase conjugate of the object beam. This real image of the object can be obtained at a distance $z_2 = (z - z_1)$ in the plane $z_1 = z$. The field distribution of real image can be represented by,

$$A_R(x_2, y_2, 2z_2 - z_1) = C_2 A_O^*(x_2 = x + 2(z_2 - z)sin\theta, y_2 = 0, z_1 = 2z_2 - z) \quad (1.41)$$

The Eqn.1.41 represents the real image of hologram and can be viewed from behind the hologram and one can see 3D image of object representing exact replica of origial object. The constant C_2 is complex and the real image is centered around an off-axis point represented by $(x_2 = 2(z_2 - z)sin\theta, y_2 = 0, z = 2z_2)$. In fact the reconstruction wave which form real image of hologram at exact position of original object is nothing but phase conjugate of reconstruction wave which was used to record the hologram. So, before non-linear optical concepts, phase conjugate beam was existing in holography. Thus in holographic reconstruction the beam $A^*(x, y, z)$ becomes phase conjugate of original object beam $A(x, y, z)$.

1.6 HOLOGRAPHIC MAGNIFICATIONS

The most significant property of holographic images are regarding the magnifications and de-magnifications, which can be achieved either by changing reconstructing wavelength or by using diverging spherical or converging beams. Consider Fig. 1.12 in which a point object $S_1(x_o, y_o, z_o)$ is kept at a distance l_o from the holographic plate and $S_R((x_r, y_r, z_r)$ located at a distance l_r from the holographic plate acts as point source for diverging spherical reference beam. We assume that when these two points S_1 and S_R are illuminated by a coherent laser beam then they generate two diverging spherical beams representing object and reference beams respectively. Then the amplitude distribution of both object and reference beams at the photographic plate is given by,

$$A_h(x, y, z_0) = \frac{A_o}{lo} e^{ik_1[l_o + \frac{(x_1 - x_o)^2 + (y_1 - y_o)^2}{2l_o}]} + \frac{A_r}{l_r} e^{ik_1[l_r + \frac{(x_1 - x_r)^2 + (y_1 - y_r)^2}{2l_r}]} \quad (1.42)$$

and the intensity distribution at the holographic recording plane will be,

$$I_h(x_1, y_1, z_0) = \frac{|A_o|^2}{l_o^2} + \frac{|A_r|^2}{l_r^2} + \frac{A_o A_r^*}{l_o l_r} e^{ik_1(l_o - l_r)}$$
$$e^{ik_1[\frac{(x_1 - x_o)^2 + (y_1 - y_o)^2}{2l_o} - \frac{(x_1 - x_r)^2 + (y_1 - y_r)^2}{2l_r}]}$$
$$+ \frac{A_o^* A_r}{l_o l_r} e^{-ik_1(l_o - l_r)} e^{-ik_1[\frac{(x_1 - x_o)^2 + (y_1 - y_o)^2}{2l_o} - \frac{(x_1 - x_r)^2 + (y_1 - y_r)^2}{2l_r}]} \quad (1.43)$$

where in Eqn.1.43, $l_o = (z_0 - z_o), l_r = (z_0 - z_r)$ with A_o,A_r represent complex amplitudes of object and reference beams respectively from S_1 and S_R. The recorded holographic plate after processing will give linear amplitude transmittance and that

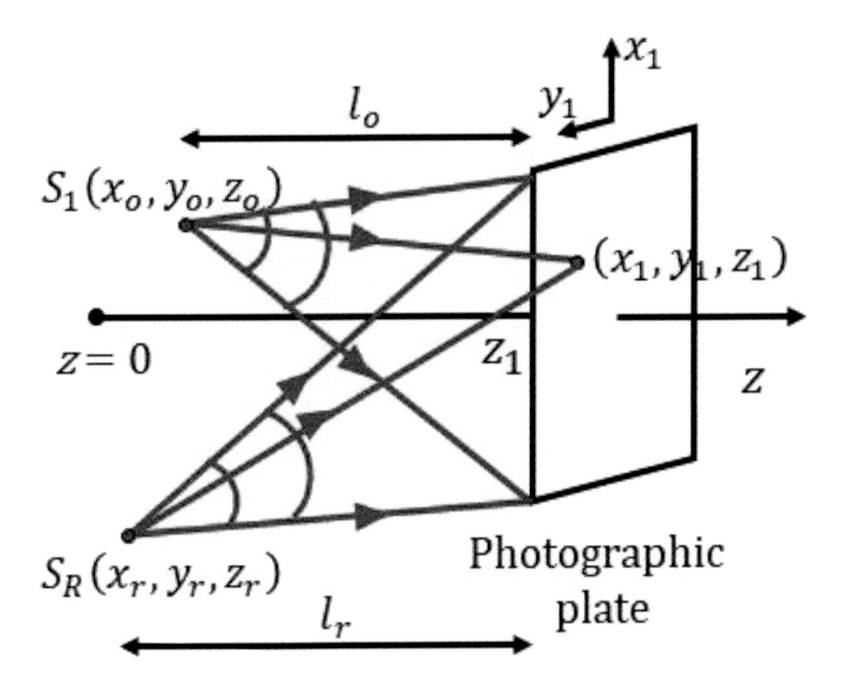

Figure 1.12 Experimental geometry for recording a hologram of point object using a spherical or divergent beam.

can be represented by,

$$A_T(x_1, y_1, z_0) = A_0 + B_0 e^{ik_1[\frac{(x_1-x_o)^2+(y_1-y_o)^2}{2l_o} - \frac{(x_1-x_r)^2+(y_1-y_r)^2}{2l_r}]} + B_0^* e^{-ik_1[\frac{(x_1-x_o)^2+(y_1-y_o)^2}{2l_o} - \frac{(x_1-x_r)^2+(y_1-y_r)^2}{2l_r}]} \tag{1.44}$$

The terms A_0, B_0 are constants and if this hologram is transilluminated by a spherical wave(Fig. 1.13) with different wavelength λ_2 and amplitude A_2 then, the field distribution at a distance Z_2 behind the holographic plate is given by,

$$\begin{aligned} A_H(x_3, y_3, z_3) = {} & CA_0 \int\int_{-\infty}^{+\infty} e^{ik_2[\frac{(x_2-x_1)^2+(y_2-y_1)^2}{2l_1} + \frac{(x_3-x_2)^2+(y_3-y_2)^2}{2z_2}]} dx_2 dy_2 \\ & + CB_0 \int\int_{-\infty}^{+\infty} e^{ik_2[\frac{(x_2-x_1)^2+(y_2-y_1)^2}{2l_1} + \frac{(x_3-x_2)^2+(y_3-y_2)^2}{2z_2}]} \\ & e^{ik_1[\frac{(x_1-x_o)^2+(y_1-y_o)^2}{2l_o} - \frac{(x_1-x_r)^2+(y_1-y_r)^2}{2l_r}]} dx_2 dy_2 \\ & + CB_0^* \int\int_{-\infty}^{+\infty} e^{ik_2[\frac{(x_2-x_1)^2+(y_2-y_1)^2}{2l_1} + \frac{(x_3-x_2)^2+(y_3-y_2)^2}{2z_2}]} \\ & e^{-ik_1[\frac{(x_1-x_o)^2+(y_1-y_o)^2}{2l_o} - \frac{(x_1-x_r)^2+(y_1-y_r)^2}{2l_r}]} dx_2 dy_2 \end{aligned} \tag{1.45}$$

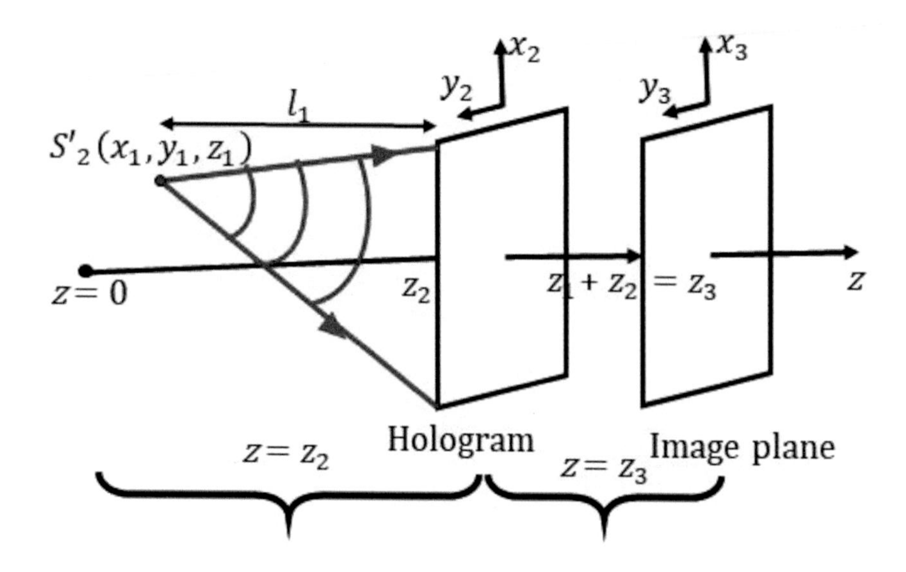

Figure 1.13 Reconstruction of a point object using spherical wavefront.

$$
\begin{aligned}
&= CA_0 e^{ik_2[\frac{(x_1^2+y_1^2)}{2l_1}+\frac{(x_3^2+y_3^2)}{2z_2}]} \times \int\int_{-\infty}^{+\infty} e^{\frac{ik_2}{2}(\frac{1}{l_1}+\frac{1}{z_2})(x_2^2+y_2^2)} \\
&\qquad e^{-ik_2[x_2(\frac{x_1}{l_1}+\frac{x_3}{z_2})+y_2(\frac{y_1}{l_1}+\frac{y_3}{z_2})]} dx_2 dy_2 \\
&+CB_0 e^{ik_1[\frac{(x_0^2+y_0^2)}{2l_0}-\frac{(x_r^2+y_r^2)}{2l_r}]} e^{ik_2[\frac{(x_1^2+y_1^2)}{2l_1}+\frac{(x_3^2+y_3^2)}{2z_2}]} \\
&\qquad \times \int\int_{-\infty}^{+\infty} e^{\frac{i}{2}[\frac{k_1}{l_0}-\frac{k_1}{l_r}+\frac{k_2}{l_1}+\frac{k_2}{z_2}](x_2^2+y_2^2)} \\
&\qquad e^{-ix_2(k_1\frac{x_0}{l_0}-k_1\frac{x_r}{l_r}+k_2\frac{x_1}{l_1}+k_2\frac{x_3}{z_2})} \\
&\qquad e^{-iy_2(k_1\frac{y_0}{l_0}-k_1\frac{y_r}{l_r}+k_2\frac{y_1}{l_1}+k_2\frac{y_3}{z_2})} dx_2 dy_2 \\
&CB_0^* e^{-ik_1[\frac{(x_0^2+y_0^2)}{2l_0}-\frac{(x_r^2+y_r^2)}{2l_r}]} e^{ik_2[\frac{(x_1^2+y_1^2)}{2l_1}+\frac{(x_3^2+y_3^2)}{2z_2}]} \\
&\qquad \times \int\int_{-\infty}^{+\infty} e^{\frac{-i}{2}[\frac{k_1}{l_0}-\frac{k_1}{l_r}-\frac{k_2}{l_1}-\frac{k_2}{z_2}](x_2^2+y_2^2)} \\
&\qquad e^{-ix_2(-k_1\frac{x_0}{l_0}+k_1\frac{x_r}{l_r}+k_2\frac{x_1}{l_1}+k_2\frac{x_3}{z_2})} \\
&\qquad e^{-iy_2(-k_1\frac{y_0}{l_0}+k_1\frac{y_r}{l_r}+k_2\frac{y_1}{l_1}+k_2\frac{y_3}{z_2})} dx_2 dy_2
\end{aligned}
\tag{1.46}
$$

In Eqn.1.46 the value of constant $C = \frac{-i}{\lambda_2}\frac{A_1}{l_1}\frac{e^{ik_2(l_1+z_2)}}{z_2}$. In Eqn.1.46, the first term represents zeroth order in diffracted beam while reconstructing and the second term is the virtual image provided the co-efficient of term $(x_2^2+y_2^2)$ i.e $[\frac{k_1}{l_0}-\frac{k_1}{l_r}+\frac{k_2}{l_1}+\frac{k_2}{z_2}]=0$.

This gives,

$$\frac{k_2}{l_v} = \frac{-k_1}{l_0} + \frac{k_1}{l_r} - \frac{k_2}{l_1} \quad (1.47)$$

Applying above condition to second term of Eqn.1.46, we get for amplitude of virtual image of the object as,

$$E_v(x_3, y_3, z_0 + l_v) = C_v \delta[(k_1 \frac{x_0}{l_0} - k_1 \frac{x_r}{l_r} + k_2 \frac{x_1}{l_1} + k_2 \frac{x_3}{l_v}) + (k_1 \frac{y_0}{l_0} - k_1 \frac{y_r}{l_r} + k_2 \frac{y_1}{l_1} + k_2 \frac{y_3}{l_v})] \quad (1.48)$$

In Eqn.1.48 the term C_v represents a suitable constant. The distance of virtual image from hologram can be obtained from Eqn.1.47,

$$\frac{1}{l_v} = \frac{\lambda_2}{\lambda_1}(\frac{1}{l_r} - \frac{1}{l_o}) - \frac{1}{l_1} \quad (1.49)$$

The Eqn.1.49 can be simplified to find value of l_v as,

$$l_v = \frac{\lambda_1 l_o l_r l_1}{\lambda_2 l_o l_1 - \lambda_2 l_r l_1 - \lambda_1 l_o l_r} \quad (1.50)$$

From Eqn.1.50 it is clear that one can change the position of virtual image by changing positions and wavelengths of reconstruction beams. If the value of $l_r = l_1$ i.e the distance of reference beam from source to the hologram plane is equal to distance of recosntructing beam from source then Eqn 1.41 reduces to $l_v = -l_o$. This shows that the virtual image of hologram is formed at same distance as the location of object. Negative sign shows that it is virtual image similar to a convex lens.

1.6.1 LATERAL MAGNIFICATIONS

The lateral magnification of virtual images along x, y directions i.e M_x^v and M_y^v are given by,

$$k_2 \frac{x_3}{l_v} = -k_1 \frac{x_0}{l_0} + k_1 \frac{x_r}{l_r} - k_2 \frac{x_1}{l_1} \quad (1.51)$$

and along y-direction the lateral magnification is,

$$k_2 \frac{y_3}{l_v} = -k_1 \frac{y_0}{l_0} + k_1 \frac{y_r}{l_r} - k_2 \frac{y_1}{l_1} \quad (1.52)$$

Further, if Δx_3 and Δy_3 represent small lateral displacements of virtual image of point objects Δx_o and Δy_o respectively then,

$$M_x^v = \frac{\Delta x_3}{\Delta x_o} = M_y^v = \frac{\Delta y_3}{\Delta y_o} = \frac{-\lambda_2 l_v}{\lambda_1 l_v} = \frac{1}{1 - \frac{l_o}{lr} + \frac{\lambda_1 l_o}{\lambda_2 l_1}} \quad (1.53)$$

Similarly, the lateral magnifications of real image can be obtained from field distribution from the third term in Eqn. 1.46 as,

$$E_{re}(x_3, y_3, z_0 + l_{re}) = C_r\delta[-k_1\frac{x_0}{l_0} + k_1\frac{x_r}{l_r} + k_2\frac{x_1}{l_1} + k_2\frac{x_3}{l_{re}}, \\ -k_1\frac{y_0}{l_0} + k_1\frac{y_r}{l_r} + k_2\frac{y_1}{l_1} + k_2\frac{y_3}{l_{re}}], \tag{1.54}$$

The term C_r represents constant term and for locating real image we have,

$$\frac{k_2}{l_{re}} = \frac{k_1}{l_o} - \frac{k_1}{l_r} - \frac{k_2}{l_1} \tag{1.55}$$

$$l_{re} = \frac{\lambda_1 l_o l_r l_1}{\lambda_2 l_o l_1 - \lambda_2 l_o l_1 - \lambda_1 l_o l_r} \tag{1.56}$$

Then lateral magnifications for real image are given by,

$$M_x^{re} = M_y^{re} = \frac{1}{1 - \frac{l_0}{l_r} - \frac{\lambda_1 l_o}{\lambda_2 l_1}} \tag{1.57}$$

It is clear from Eqn.1.57 and 1.53 that the magnifications of real and virtual images are different.

1.6.2 LONGITUDINAL MAGNIFICATIONS

The longitudinal magnifications of real and virtual images are obtained by differentiating 1.50 and 1.56 and we get,

$$M_z^v = \frac{-dl_v}{dl_o} = \frac{\lambda_1 l_r^2 l_1^2}{(\lambda_2 l_o l_1 - \lambda_2 l_r l_1 - \lambda_1 l_o l_r)^2} \tag{1.58}$$

$$M_z^{re} = \frac{dl_{re}}{dl_o} = \frac{\lambda_1 l_r^2 l_1^2}{(\lambda_2 l_o l_1 - \lambda_2 l_o l_1 - \lambda_1 l_o l_r)^2} \tag{1.59}$$

The relationship between lateral and longitudinal magnifications can be obtained from Eqns.1.53, 1.57, 1.58 and 1.59 as,

$$M_z^v = \frac{\lambda_1}{\lambda_2}(M_x^v)^2 = \frac{\lambda_1}{\lambda_2}(M_y^v)^2 \tag{1.60}$$

$$M_z^{re} = \frac{\lambda_1}{\lambda_2}(M_x^{re})^2 = \frac{\lambda_1}{\lambda_2}(M_y^{re})^2 \tag{1.61}$$

If the reference beam ($l_r = \infty$) and reconstruction beam ($l_1 = \infty$) are collimated then $M_z = M_x = M_y = 1$.

1.7 REFLECTION HOLOGRAPHY

1.7.1 CONSTRUCTION

The reflection holography or white light holography is the basic recording procedure for obtaining rainbow holography(Appendix B) where the construction of hologram is done using laser beam and reconstruction is carried out using white light. This is the principle behind construction of rainbow holograms which are widely used in advertisements. The experimental geometry for recording a reflection hologram of a

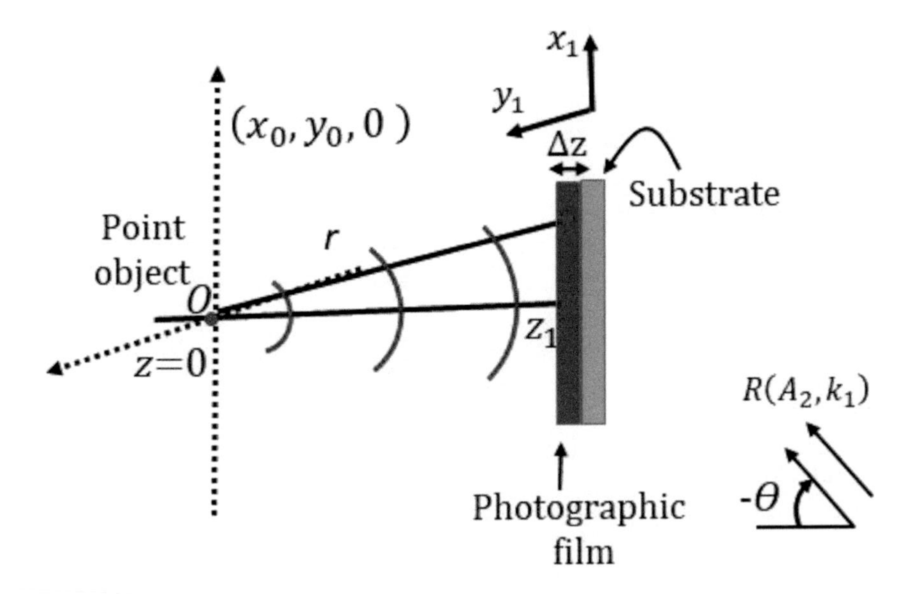

Figure 1.14 Recording of reflection hologram of a point object.

point objecrt is shown in Fig.1.14. Unlike conventional holographic recording plate, reflection type holographic recording requires thick silver bromide emulsion(Film of holographic plate) over glass substrate. Consider Fig, 1.14 in which a point object is located at a point P on-axis($z = 0$) and the holographic plate with thickness of silver bromide emulsion layer ΔZ is located at a distance z_o. Now the field distrbution from the holographic plate located at a plane $(z_0 + z_1)$ is given by,

$$A(x_1, y_1, z_o + z_1) = A_r e^{i\beta_1[-(z_o+z_1)cos\theta + x_1 sin\theta]} + \frac{A_o}{z_o} e^{i\beta_1[z_o+z_1+\frac{(x_1^2+y_1^2)}{2(z_o+z_1)}]} \quad (1.62)$$

where $0 \leq z_1 \leq \Delta z$,and A_r and A_o are the amplitudes of reference and object beams(spherical beam originating from the point object) respectively.

Approximating the value of R as,

$$r = [z_o + z_1 + \frac{(x_1^2 + y_1^2)}{2(z_o + z_1)}] \tag{1.63}$$

Then the intensity of obejct and reference beams at the plane of holographic plate is,

$$I(x_1, y_1, z_o + z_1) = |A_r|^2 + |\frac{A_o}{z_o}|^2 + \frac{A_r^* A_o}{z_o}$$
$$e^{i\beta_1[(z_o+z_1)(1+cos\theta) - x_1 sin\theta + \frac{(x_1^2+y_1^2)}{2(z_o+z_1}]}$$
$$\frac{A_r A_o^*}{z_o} e^{-i\beta_1[(z_o+z_1)(1+cos\theta) - x_1 sin\theta + \frac{(x_1^2+y_1^2)}{2(z_o+z_1}]} \tag{1.64}$$

Simplifying further we get,

$$= |A_r|^2 + |\frac{A_o}{z_o}|^2 + \frac{2}{z_o}|A_o||A_r| cos\beta_1[(z_o + z_1)]$$
$$\times(1 + cos\theta) - x sin\theta + \frac{(x_1^2 + y_1^2)}{2(z_o + z_1)} + \phi_o] \tag{1.65}$$

The term ϕ_o is the phase difference between the object and reference beams. Eqn.1.65 represents intensity distribution in different planes of silver bromide emulsion of the holographic plate. Since the interefrence pattern between object and reference beams have standing wave character the intensity or silver bromide emulsion shows a periodic variation with distnace z_1. Thus the planes in the holographic plate with similar exposures will lie at a distance $\frac{\lambda}{2}$ to each other. The reconstruction of a reflection hologram is similar to X-ray diffraction by certain crystals[11]. This is because the reflection hologram is volume hologram type, where thickness of silver bromide film is higher than compared to a transmission hologram. The amplitude reflectance from a reflection hologram is,

$$E_{Rh}(x_1, y_1, z_o + z_1) = A_o + B_o e^{i\beta_1[(z_o+z_1)(1+cos\theta) - x_1 sin\theta + \frac{(x_1^2+y_1^2)}{2(z_o+z_1)}]}$$
$$+ B_o^* e^{-i\beta_1[(z_o+z_1)(1+cos\theta) - x_1 sin\theta + \frac{(x_1^2+y_1^2)}{2(z_o+z_1)}]} \tag{1.66}$$

with A_o and B_o are usual constants.

1.7.2 RECONSTRUCTION

The recosntruction geometry for a reflection hologram is shown in Fig. 1.15. A collimated beam is used for the reconstruction and if it is white light then the respective spectrum of white light i.e if construction of hologram was done in 6328 A He-Ne laser then only that wavelength will satisfy the Bragg condition($2dsin\alpha = n\lambda$).

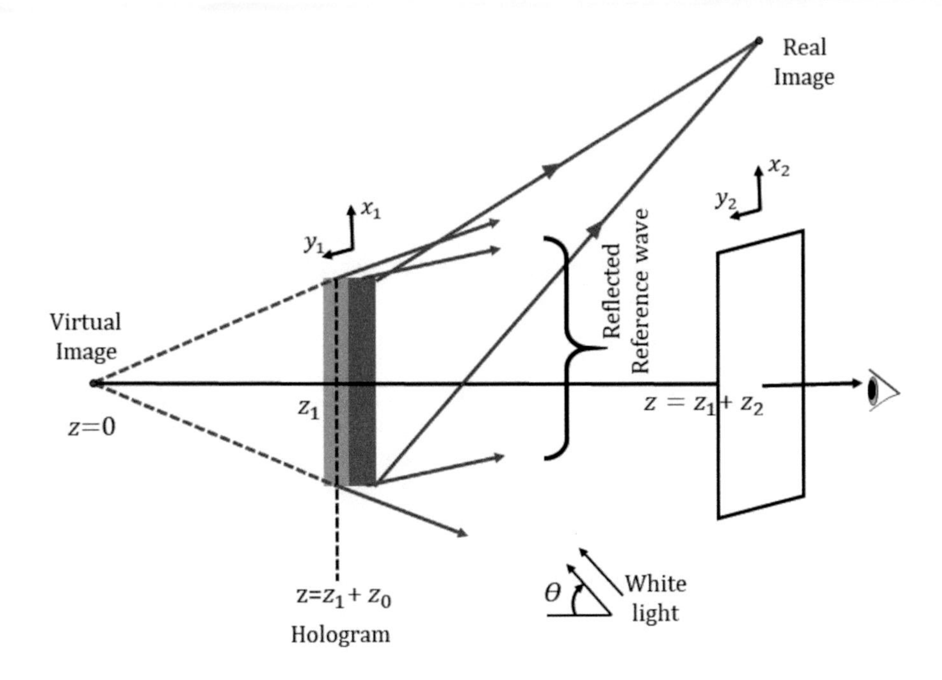

Figure 1.15 Reconstruction from a reflection hologram of a point object using white light.

Now the field distribution of reflected light beam within the layer $z = z_o + z_1$ is given by,

$$E_H(x_1, y_1, z_o + z_1) = Ae^{-i\beta_1[(z_o+z_1)(cos\theta)-x_1 sin\theta]}E_{Rh}(x_1, y_1, z_o + z_1) \tag{1.67}$$

and the field distribution of reflected light will be obtained at the plane $z = z_o + z_1 + z_2$ as,

$$\begin{aligned} E_H(x_2, y_2, z_o + z_1 + z_2) = \frac{-i}{\lambda_1}\frac{e^{i\beta_1 z_2}}{z_2}Ae^{-i\beta_1[(z_o+z_1)(cos\theta)-x_1 sin\theta]} \\ \times A_o + B_o e^{i\beta_1[(z_o+z_1)(1+cos\theta)-x_1 sin\theta+\frac{(x_1^2+y_1^2)}{2(z_o+z_1)}]} \\ +B_o^* e^{-i\beta_1[(z_o+z_1)(1+cos\theta)-x_1 sin\theta+\frac{(x_1^2+y_1^2)}{2(z_o+z_1)}]} \\ \times e^{\frac{i\beta_1}{2z_2}[(x_2-x)^2+(y_2-y)^2]}dx_1 dy_1 \end{aligned} \tag{1.68}$$

If the value of $z_2 = z_o + z_1$ then Eqn.1.59 reduces to,

$$\begin{aligned} E_H(x_2, y_2, 2(z_o + z_1)) = C_1(z_1)e^{i\beta_1 x_2 \sin\theta} + C_2(z_1)e^{\frac{i\beta_1}{4(z_o+z_1)}}(x_2^2 + y_2^2) \\ +C_3(z_1)\delta(x_2 - 2(z_o + z_1)sin\theta, y_2) \end{aligned} \tag{1.69}$$

The first term in Eqn.1.69 is the reflected white light component by the hologram layer at $z = z_o + z_1$ for $\beta = \beta_1$ and C_1, C_2, C_3 represent complex constants. The second term is the virtual image of the object located at exact position of the original position of the point object. Also, all layers in the volume hologram produce virtual images at exactly same position of the orignal point object. The third term in Eqn 1.69 represents real image of the point object at an off-axis position $x_2 = 2(z_o + z_1)sin\theta, y_2 = 0$. Fig. 1.15 shows all the reflected field distribution. The net optical field distribution of virtual image is nothing but the superposition of optical fields by different layers of hologram and that is equal to,

$$E_v(z=0) = \int_0^{\Delta z} C_2(z_1)dz_1 \tag{1.70}$$

Unlike transmission holography, the reflection holograms can be recorded with two or more coherent laser beams of different wavelengths, simultaneously and the reference beam also should have same coherence. The resultant optical field distribution in any silver bromide emuslion layer will be,

$$E(x_1, y_1, z_o + z_1) = \Sigma_j [A_{rj} e^{i\beta_j[-(z_o+z_1)cos\theta + x_1 sin\theta]} + \frac{A_{oj}}{z_o} e^{i\beta_j[z_o+z_1+\frac{(x_1^2+y_1^2)}{2(z_o+z_1)}]}] \tag{1.71}$$

The constants A_{rj}, A_{oj} represent amplitudes of reference and object beams respectively and they correspond to respective wave length λ_j. Now, the amplitude reflectance of this reflective hologram is given by,

$$E_R(x_1, y_1 z_o + z_1) = A_{r_0} + \Sigma_j [B_{j_0} e^{\frac{i\beta_j}{2(z_o+z_1)}[x_1^2+y_1^2-2(z_o+z_1)x_1 sin\theta]} + B^*_{j_c} e^{-\frac{i\beta_j}{2(z_o+z_1)}[x_1^2+y_1^2-2(z_o+z_1)x_1 sin\theta]}] \tag{1.72}$$

The constants $A_{r_0}, B_{j_o}, B_{j_c^*}$ in Eqn.1.72 represents respective complex constants and when this reflection hologram is illuminated by a white light, then multicolor images are generated in reflected light.

This reflection hologram is actually the master hologram for creating multiple rainbow holograms. The reflection hologram requires to satisfy Bragg condition that is the direction of reconstruction wave must coincide with direction of reference wave during hologram recording. One can store several images of different objects in same holographic plate by changing the directions of reference waves for each image and can be reconstructed by changing direction of reconstructing waves. This property of reflection hologram makes it possible to use as a holographic memory and such type of memory is known as ROM(Read only Memory).

1.8 PRACTICAL DEMONSTRATION OF HOLOGRAPHY

Due to advancements of recording materials especially photopolymers where one can directly record a beautiful 3D object image in both reflection and transmission modes. The photopolymer used in this experiment was LLPF465 supplied by Ms. Light Logics, Thiruvallom, Trivandrum, India and is an excellent holographic

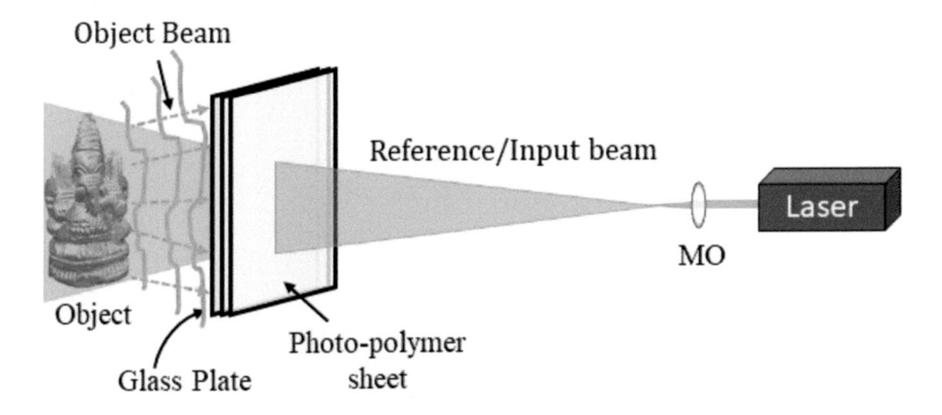

Figure 1.16 Recording of a conventional reflection hologram of idol of Lord Ganesha.

recording medium for both reflection and transmission type volume holograms. This holographic photopolymer film LLPF465 has i) High Diffraction Efficiency ii) High Refractive Index Modulation iii) No need for wet processing like Agfa Gaevert 10E75 plates iv) It can work under relatively bright safe light v) Easy to cut and handle and also cost effective. The photopolymer thin film LLPF465 of 15 microns thickness is laminated on to a glass substrate. Fig. 1.16 shows a typical experimental geometry where light beam from a green Nd:Yag (140 mW) laser having wavelength 532 nm, passes though the polymer thin film sheet and falls on a 3D object(Lord Ganesha) and the reflected light from the object falls on photopolymer sheet as shown in Fig. 1.14. After few seconds of exposure the photopolymer based holographic film is bleached using a constant illumination of mercury white light as the bleaching process. Alternatively, the hologram can also be exposed to sunlight for about 5 minutes. Fig. 1.17 shows the hologram of Lord Ganesha when

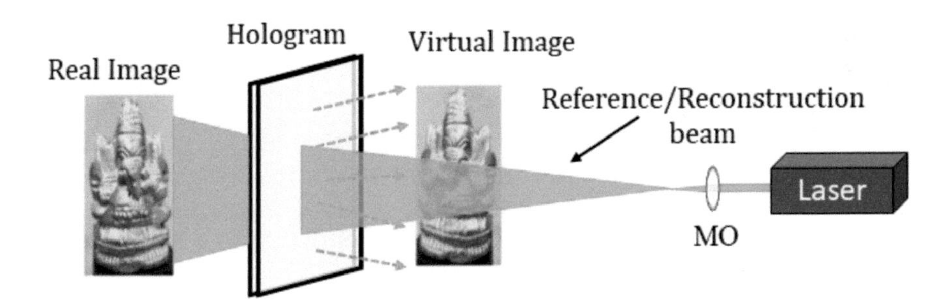

Figure 1.17 Reconstruction of the hologram of idol of Lord Ganesha.

Figure 1.18 Reconstruction of the hologram of idol Lord Ganesha using white light.

the recorded photopolymer hologram was illuminated by same laser beam which recorded it, but one can reconstruct the holographic image using even white light from a mobile phone. Fig. 1.18 shows holographic image of idol Lord Ganesha when the photopolymer film based hologram was illuminated by a white light and is classical example of relection hologram which we had discussed earlier in this chapter.

2 Conoscopic Holography

2.1 INTRODUCTION

So far we have discussed regarding holographic imaging only using coherent laser sources. But, simulatanously efforts are also made by researchers to develop holography using incoherent or partially coherent sources. In conventional holographic recording and reconstruction using laser beams, each object point, the intensity and the lateral and longitudinal locations of the object, is interferometrically recorded as a Fresnel zone plate (FZP) and stored on a holographic films like Agfa Gaevert/Kodak films. There are several experimental techniques have been proposed for incoherent recording, such as i) shadow casting' and ii) interferometric systems including an amplitude-splitting interferometer based on a double-focus birefringent lens. Main problems in these incoherent holographic techniques are due to need of severe mechanical and electronics requirements along with low signal-to-bias ratio. One of the methods, to develop incoherent holography was by Sirat and Psaltis[12-16] using a birefringent crystal and they named it as conoscopic holography. Whenever light propagates through a birefringent crystal the incident light is split in to two as ordinary and extraordinary beams. These beams combine while coming out of the crystal and an analyzer kept just outside the birefringent crystal produces the interference between the ordinary and the extraordinary beams. This interference phenomena occurs due to phase difference introduced by the extraordinary index of refraction angle-dependent change in optical path inside the birefringent crystal. The birefringent phenomena occuring inside a uniaxial crystal is responsible for the formation of the conoscopic figures. In conoscopic holography this birefringent effect of uniaxial crystal is used and the most advantageous feature of conoscopic incoherent holography is that the two interfering light beams have identical geometrical paths[12]. Also, the natural space invariance of the system permits both ordinary and extraordinary beams to have equal optical paths over the full image frame, and due to that the spatial coherence of the light source imposes no limitations on the sizes of the object and the hologram.

2.2 CONSTRUCTION OF CONOSCOPIC HOLOGRAPHY

Consider Fig. 2.1 in which the experimental geometry for recording conoscopic hologram is given.The incoherent and un-polarized light emerging from a point object passes through a circular polarizer which, generates two orthogonal polarizations states with one of them having a quarter-wavelength relative phase delay. The two beams then propagate through a uniaxial crystal (shown in Fig. 2.1) along ordinary and extraordinary ray path inside the uniaxial crystal. These two beams travel with different velocities inside the crystal due to birefringent effect with refractive index of ordinary ray and for the extraordinary ray given by n_1 and n_2 respectively. The

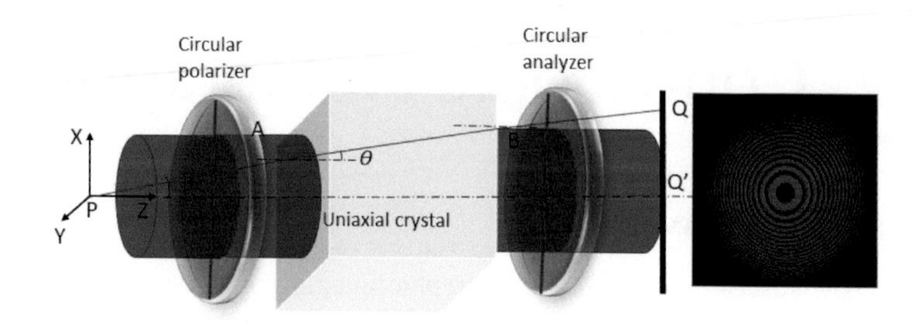

Figure 2.1 Recording geometry of a conoscopic holography using a uniaxial crystal.

ordinary ray refractive index is constant unlike the extraordinary refractive index. Thus the phase delay between ordinary ray and extraordinary ray is a function of the direction of the ray relative to the crystal optical axis. These two rays combine at the output of uniaxial crystal and then passes through the circular analyzer placed after the crystal. The analyzer makes these two rays in to same polarization state and due to that they interfere. Now, like conventional interference the constructive and destructive ineterference will occur depending upon phase delay between the ordinary and the extraordinary rays. Further the circular analyzer compensates for the initial quarter-wavelength delay. Let a point source $P(x,y,z)$ which is kept infront of conocopic holographic set up as shown in Fig. 2.1 emits a quasi monochromatic light with a wavelength λ. The light beam then passes through the circular polarizer and uniaxial crystal where it is split in to two one being ordinary and other being extra ordinary ray respectively. If $E_p(t)$ represents amplitude distribution of light field at the source plane $P(x,y,z)$ then this field passes through a circular polarizer the uniaxial crystal. Inside uniaxial crystal the incident beam from source P is split in to ordinary beam with field amplitude $E_o(t)$ and extraordinary beam with field amplitude $E_e(t)$ respectively. They recombine just outside the uniaxial crystal and passes through an analyzer where both (ordinary and extraordinary)beams will be in same polarization state. Now at the detector plane Q The amplitude distribution functions of ordinary and extraordinary rays are given by,

$$E_o(Q,t) = \frac{a_o}{j\lambda l_o} u(P,t + \frac{l_o(t)}{c}) \tag{2.1}$$

$$E_e(Q,t) = \frac{a_e}{j\lambda l_e} u(P,t + \frac{l_e(t)}{c}) \tag{2.2}$$

where, a_o, a_e are amplitude transmittances of ordinary and extraordinary rays, respectively, and the terms in denominators of Eqns 2.1 and 2.2 represent Fresnel-Kirchoff propagation characteristics and l_o, l_e represent the distances travelled by ordinary and extraordinary rays inside the uniaxial crystal of length l. In most cases the amplitude transmittances are considered to be equal but there will be slight differences introduced by i) Fresnel's (Snell's) laws in their values due to the differences in transmission coefficients of the ordinary and the extraordinary beams ii) the discrepancies in amplitudes that are due to the slightly different angles of emergence of the ordinary and the extraordinary beams and finally iii)there may be selective absorption with respect to polarization in one of the elements of the system.

2.2.1 THEORETICAL EXPLANATION

Consider Fig. 2.1 in which the ray travels from source point P to detector point Q via birefringent uniaxial crystal of length l. The initial ray coming from object source point will travel with different velocities in the crystal due to refractive indices n_o and n_e for ordinary and extraordinary rays respectively.The phase delay introduced due to the birefrinegnt effect in the optical path length between the ordinary and the extraordinary rays is given by following[12-16],

$$l_o = \vec{PA} + n_o\vec{AB} + \vec{BQ}, \tag{2.3}$$

$$l_e = \vec{PA} + n_e\vec{AB} + \vec{BQ} \tag{2.4}$$

Further assuming that crystal length is l which is equalent to AB and neglecting the influence of the non-normal incidence of the rays and the effects that are due to crystal optics on the extraordinary ray, the difference between $(l_e - l_o)$ yields,

$$n_e(\theta) - n_o(l) = \Delta n_e(\theta)l, \qquad \Delta n_e(\theta) = \Delta n\frac{n_o^2}{n_e^2}\theta^2 \tag{2.5}$$

where the θ represents angle direction of the ray inside the crystal and is equivalent to $\theta = \frac{r}{z_c}$, with $r = \sqrt{(x-x')^2 + (y-y')^2}$. Then the phase difference $\Delta\phi$ value becomes,

$$\Delta\phi = 2\pi\frac{l_o - l_e}{\lambda} = 2\pi\frac{\Delta n l r^2}{\lambda n_e^2 z_c^2} = \pi\frac{\kappa_0 r^2}{z_c^2} \tag{2.6}$$

In Eqn. 2.6 $\Delta n = (n_o - n_e$ is birefringence value, l is birefringence crystal length and $z_c = (z - z')$. Now, the total intensity of light beam at the detector for positive conoscopic holography will be,

$$I_+(Q,t) = \frac{\varepsilon_0 c}{2}[(\langle E_o(Q,t) + E_e(Q,t)\rangle\rangle)(\langle E_o^*(Q,t) + E_e^*(Q,t)\rangle\rangle)] \tag{2.7}$$

We can consider that the transmission coefficients of the ordinary and the extraordinary beams are al- most equal and because of that i) The differences in transmission coefficients of the ordinary and the extraordinary beams that are due to Fresnel's (Snell's) laws are neglible and ii) The discrepancies in amplitudes that are due to the slightly different angles of emergence of the ordinary and the extraordinary beams, can also be neglected and finally iii) any selective absorption with respect to polarization in one of the elements of system also can be ignored. Defining bias (a_b) and signal(a_s) amplitudes transmittances as,

$$a_b^2 = a_o^2 + a_e^2 \tag{2.8}$$

$$a_s^2 = 2a_e a_o \tag{2.9}$$

and since, the amplitude transmittances a_e, a_o are angle dependent, a_b, a_s are also angle dependent. We have to normalize the size of pinhole at point P and if the size of it is Δs, then

$$\int_{\Delta s} \frac{ds}{\lambda^2} = 1 \tag{2.10}$$

This assumption will take care of the wavelength dependence in the denominators of the ordinary and the extra-ordinary amplitude distribution functions of Equations 2.1, 2.2, respectively . For other emitting-point sizes the intensities at the detector will be scaled linearly with the ratio of the surface areas. Now, for the positive on-axis consocopic hologram we get,

$$I_+(Q,t) = (\frac{\varepsilon_0 c}{2l_o^2}[(a_o^2 + a_e^2)\langle|E(P,t))|^2\rangle(2a_o a_e Re[\Gamma(P, \frac{l_o - l_e}{c}]) \tag{2.11}$$

Using bias and signal transmission co-efficients a_b, a_s values from Eqns.2.7, 2.8 we get,

$$I_+(Q,t) = \frac{1}{l_0^2}[a_b^2 + a_s^2 Re[\gamma(P, \frac{l_o - l_e}{c}]]I(P) \tag{2.12}$$

Similarly for the negative on-axis conoscopic holography the detected intensity of light beam is given by,

$$I_-(Q,t) = \frac{\varepsilon_0 c}{2}[(\langle E_o(Q,t) - E_e(Q,t))\rangle)(\langle E_o^*(Q,t) - E_e^*(Q,t))\rangle)] \tag{2.13}$$

Substituting respective values we get,

$$I_-(Q,t) = \frac{1}{l_0^2}[a_b^2 - a_s^2 Re[\gamma(P, \frac{l_o - l_e}{c}]]I(P) \tag{2.14}$$

Considering the first part of positive intensity of conoscope in Eqn. 2.12 and negative intensity of conoscope in Eqn. 2.14 respectively, and combining it we get,

$$I_T = \frac{I_+(Q) + I_-(Q)}{2} = \frac{a_b^2}{l_0^2} I(P) \tag{2.15}$$

The Equation 2.15 represents half the equivalent isotropic intensity. On the other hand, the second terms in positive and negative intensities of conoscope actually represent the hologram and is given by follwing expression,

$$H(Q) = \frac{I_+(Q) - I_-(Q)}{2} = \frac{a_s^2}{l_0^2} Re[\gamma(P, \frac{l_o - l_e}{c}]]I(P) \tag{2.16}$$

But, the value of $l_o = \frac{z_l}{cos\beta}$ [16] and introducing that in Eqn.2.16, we get,

$$h(M) = \frac{a_s^2 cos^2\beta}{z_l^2} Re[\gamma(P, \frac{l_o - l_e}{c})] \tag{2.17}$$

The Eqn.2.17 represents the recorded Conoscopic Point Spread Function. Introducing the phase difference between ordinary and extraordinary waves propagating through birefringent crystal as,

$$\Delta\phi = \frac{2\pi}{\lambda}\Delta l = \frac{2\pi}{\lambda}(l_e - l_0) \tag{2.18}$$

where,

$$l_e = \frac{[(z - z') - l]}{cos\beta_e} + \frac{n_e(\theta_e)l}{cos\theta_E}cos(\theta_E - \theta_e) \tag{2.19}$$

and

$$l_o = \frac{[(z - z') - l]}{cos\beta_o} + \frac{n_o l}{cos\theta_o} \tag{2.20}$$

Now, $\Delta l_t = (l_e - l_o)$

2.2.2 CONSTRUCTION OF CONOSCOPIC HOLOGRAM

As shown in Fig. 2.2 consider the point P(x,y,z) represents the location of point source and $Q(x', y', z')$ represent the location of recording position. If a light beam is passed through the conoscope then the hologram of point source will become,

$$H(x', y', z') = \int_v I(x,y,z) \times H[j\pi\kappa_0 \frac{(x - x')^2 + (y - y')^2}{(z_c - z')^2}]dxdydz \tag{2.21}$$

The term z_c is close to the longitudinal distance $z - z'$ as described in references[13,14]. The Eqn 2.17 can also be represented interms of 2 dimensional convolution as,

$$H(x', y', z') = \int_{z_c} I(x,y,z) * H[j\pi\kappa_0 \frac{(x^2 + y^2)}{(z_c - z')^2}]dz_c \tag{2.22}$$

In above analysis consider the important difference between the z depen- dences of conoscopic and conventional coherent holography where, the conoscopic point

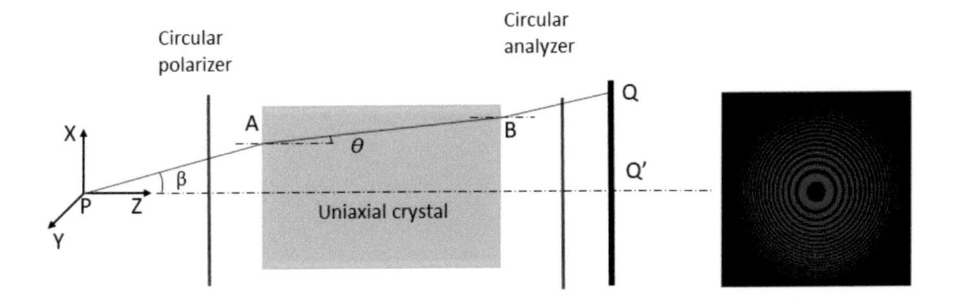

Figure 2.2 Experimental demonstration of a conoscopic hologram.

spread function is a nonsingular three-dimensional transfer function. If the crystal length had been equal to the distance between the emitting and recording points, the term $\frac{L}{z}$ would have been unity and the dependence would have been the usual $\frac{1}{z}$ dependence. On the other hand, if the crystal length is smaller or bigger than this distance, the effect is scaled by the ratio $\frac{L}{z}$, where L is the total length of birefringent crystal, representing the proportion of active media in the path of the light. Unlike conventional holography, the reconstruction of a conoscopic hologram can be done using either coherent or incoherent light. This is because the conoscopic holography is the incoherent superposition of the elementary conoscopic figures. Fig. 2.2 shows the conoscopic hologram of superposition of several point images. Though conoscopic hologram has different approach, one can apply it for wavefront sensing[17] and polarization based phase imaging for stress and strain measurements but due to developments in digital holography and for want of high quality birefringent crystals, very few conoscopic holographic imaging applications have been developed.

3 Computer-Generated holography

3.1 INTRODUCTION

A conventional hologram is produced by recording the interference pattern of an object and reference beam. In computer generated holography, the object need not exist and it is enough if the object is known in mathematical form. The idea was first proposed by Adolf Lohmann[18] and later many others[19-21] have contributed significantly to computer generated hologram. The computer-generated holograms are binary as they consist of many transparent dots on an opaque background. The binary character of computer holograms facilitate their reconstruction in real sense. To begin with the holograms are plotted in black and white on a large scale, then they are optically reduced in size and recorded on film. The transmittance of computer generated holograms can assume the values zero or one like digital, but when it is reconstructed using laser beam then the reconstructed images look very similar to conventional holograms of com- parable dimensions. There are four basic stages in the synthesis of computer-generated holograms which are i) Formulating mathematical models of the object to be designed (ii) Computing the mathematical hologram which is an array of complex numbers that represent amplitudes and phases of hologram samples in the hologram plane (iii) Encoding samples of the mathematical hologram for recording them on the physical medium by encoding and are then converted into an arrays of numbers that control optical properties of the physical recording medium used for recording the hologram. (iv)Finally, fabrication of the computer generated hologram. The computer-generated holograms can be used as spatial filters for optical information processing, as computer generated diffractive optical elements, such as beam forming elements for laser tweezers, laser focusers, deflectors, beam splitters and multiplicators, and for information display. The mathematical models of computer generated hologram describe the geometry of wave propagation from objects to the hologram plane and specify criteria for evaluation of hologram performance. In computer generated holography, the mathematical model of the object is intended to specify the amplitude and phase distribution of the object wave front. There are three types of the models which are used for generating CGH(Computer Generated Holography). One can have the analytical models, geometrical models representing objects as compositions of elementary diffracting elements such as point scatters, segments of 2-D or 3-D curves, slits,polygonal mesh or wire-frame models in 3D computer graphics "raster graphics", or "bitmap" models that represent objects in form of 2-D or 3-D arrays of points. The analytical models are used for generating diffractive optical elements. Using such analytically specified desired distribution of an object wave front, the synthetic diffractive optical element is computed using methods of analytical or numerical integration with appropriate oversampling to avoid transformation sampling artifacts. Then it is sampled to generate sampled version for recording on physical media. In other case of objects represented using geometrical models, mathematical hologram is computed as a superposition of elemental holograms which correspond to the elements such as point scatters, edges of wire frame models, faces of polygonal mesh models, etc and are stored. In the next section the mathematical model for computer generated display holograms of objects specified by "raster graphic" model is described.In monochromatic illumination, object characteristics that define its ability to reflect, transmit or scatter incident radiation can be described for our purposes by a radiation (transmission) reflection factor with respect to the light complex amplitude.

3.2 MATHEMATICAL MODEL

In order to describe the theoretical explanation for a computer generated hologram we follow Yaroslavsky [21] lecture notes. Consider Fig. 3.1 which shows a schematic diagram for the visual observation of objects. In Fig. 3.1, the observer's position with respect to the object is defined by the observation surface where the observer's eyes are situated. These set of observation positions is defined by the

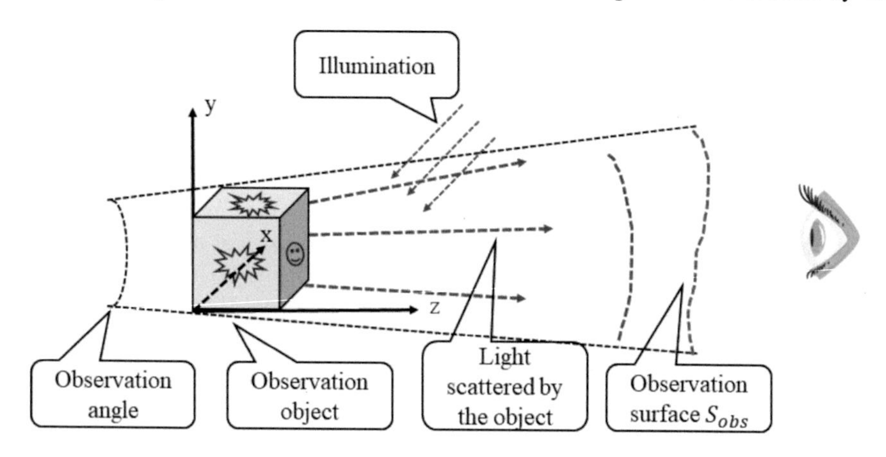

Figure 3.1 Procedure of visual observation of objects by means of computer generated holograms.

object observation angle and in order to allow the observer to see the object at the particular observation angle, it is necessary to reproduce in case of hologram, the distribution of intensities and phases of the light waves scattered by the object at the observation surface. We assume monochromatic illumination of objects for enabling to describe light wave transformations in terms of the wave amplitude and phase. The polychromatic illumination can be treated as superposition of monochromatic illumination with different wavelengths especially for generating color holograms. Considering monochromatic illumination the light complex amplitude and intensity reflected by the object beam are given by,

$$I_o(x,y,z) = |A_o(x,y,z)|^2 \tag{3.1}$$

Now, the intensity of the scattered wave $I_s(x,y,z)$ and its complex amplitude $A_s(x,y,z)$ at the point (x , y , z) are related to the intensity $I(x,y,z)$ and complex amplitude $A(x,y,z)$ of the incident wave is given by,

$$\begin{aligned} I_s(x,y,z) &= I_o(x,y,z)I(x,y,z) \\ A_s(x,y,z) &= A_o(x,y,z)A(x,y,z) \end{aligned} \tag{3.2}$$

The complex amplitude of reflected beam from the object surface $A_o(x,y,z)$ is given by,

$$A_o(x,y,z) = |A_0(x,y,z)|e^{i\theta_0(x,y,z)} \tag{3.3}$$

THe modulus term $A_0(x,y,z)$ and phase $\theta_0(x,y,z)$ show how ? the modification of complex amplitude of incident light beam on the object i.e $|A(x,y,z)|$ and the original phase $\Phi(x,y,z)$ are changed after reflection by the object surface or transmission through the point (x , y , z) on the object. Thus we obtain equations for amplitude and phase terms of object wavefront as,

$$A_{obj} = |A(x,y,z)||A_0(x,y,z)| \tag{3.4}$$

$$\Phi_{obj} = \Phi_0(x,y,z) + \theta_0(x,y,z) \tag{3.5}$$

Defining a 3D surface near the object as (ξ,ζ,η), the complex amplitude of propagating object beam at that surface can be written as,

$$A_{obs}(\xi,\zeta,\eta) = \int\int\int_{S_{obj}} A_{obj}(x,y,z)h(\xi,\zeta,\eta,x,y,z)dxdydz \tag{3.6}$$

The kernel $h(x,y,z;\xi,\eta,\zeta)$ of the wave propagation integral depends on the spatial disposition of the object and the corresponding observational surface. Now, the reconstruction of the object surface can be described by a back propagation integral over the observation surface as,

$$A^{-}_{obj}(x,y,z) = \int\int\int_{S_{obs}} A(\xi,\zeta,\eta)h^{-}(\xi,\zeta,\eta,x,y,z)d\xi d\zeta d\eta \tag{3.7}$$

In above equation, $A^{-}_{obj}(x,y,z)$ is complex amplitude of the object when viewing from the observation surface and $h^{-}(\xi,\zeta,\eta;x,y,z)$ is a kernel reciprocal to $h(\xi,\zeta,\eta;x,y,z)$. Thus, the reconstruction process of a computer generated hologram requires computation of $A(\xi,\eta,\zeta)$ through $A_{obj}(x,y,z)$, and has to be defined by the description of object surface, and illumination conditions required for recording the computation results on a physical medium in a form that enables interaction with radiation for visualizing or reconstructing the original form of object surface

$A^{-}_{obj}(x,y,z)$ as given by Eqn 3.7. The 3D integration required by Eq. 3.6 can be reduced to 2-D integration by considering the natural limitations of visual observation as i) The pupil of the observer's eye is usually much smaller than the distance to the observation surface, ii) The area of the observation surface is large compared to the inter-pupil distance, iii) The depth of objects situated at a distance convenient for visual observation usually is small compared to the distance to the observation surface. Considering all these limitations, one can consider that the observation surface con-

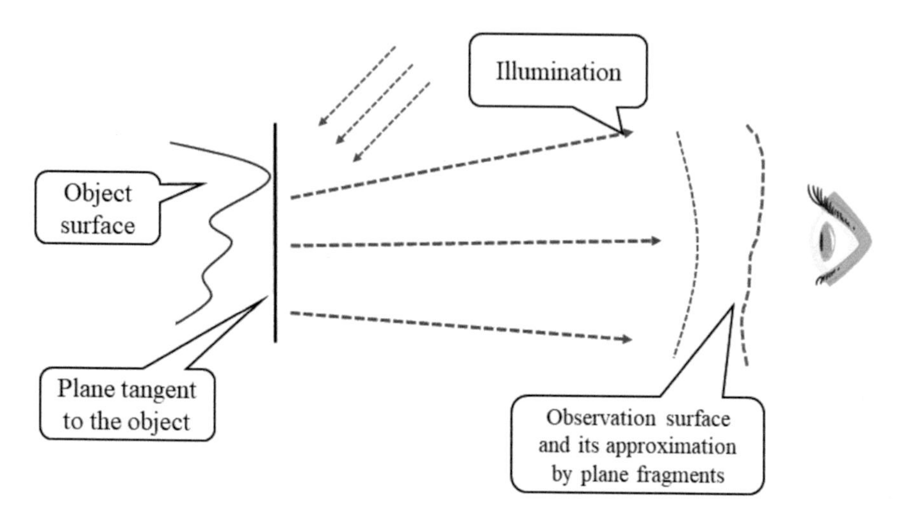

Figure 3.2 Observation geometry for computer generated hologram of an object .

sists of relatively small sub-areas approximated by planes. Further, using the laws of geometrical optics, we can replace the wave amplitude and phase distributions over the object surface by those of the wave on a plane tangent to the object and parallel to the given plane as shown in Fig. 3.2. With these considerations, the problem of hologram synthesis over the whole observation surface can be reduced to the synthesis of fragmentary holograms for the plane areas of this surface with the complete hologram being composed as a mosaic of fragmentary holograms. with these assumptions, equations for fragmentary holograms can be obtained from Eq.(3.6)in 2 dimensional form as,

$$A_I(\xi,\eta) = \int_{-\infty}^{+\infty}\int_{-\infty}^{+\infty} A_o(x,y)h_I(\xi,\eta,x,y)dxdy \tag{3.8}$$

where, $A_o(x,y) = Ao_0(x,y)e^{i\Phi_o(x,y)}$ is a complex amplitude function of light beam reflected or transmitted by the object onto the plane (x, y) and parallel to the observation plane (ξ,η) and $h_I(x,y;\xi,\eta)$ is 2-D transformation kernel for the distance between plane from (x,y) to (ξ,η) along the propagation direction z . Since in laboratory scale assuming the distance between object plane (x,y), observation plane(ξ,η)

are small compared to distance z then using Fresnel transform equation 3.8 can be simplified as,

$$A_I(\xi,\eta) = \int_{-\infty}^{+\infty}\int_{-\infty}^{+\infty} A_o(x,y)e^{\frac{i\pi}{\lambda z}(x-\xi)^2+(y-\eta)^2}dxdy \tag{3.9}$$

where λ represents wavelength of light beam used to record the computer generated hologram and further applying Fresnel approximation we get,

$$[\frac{\pi}{\lambda z}(x^2+y^2)] << 1 \tag{3.10}$$

which simplifies Eqn.3.9 as,

$$A_I(\xi,\eta) = e^{i\pi(\xi^2+\eta^2)}\int_{-\infty}^{+\infty}\int_{-\infty}^{+\infty} A_o(x,y)e^{\frac{i2\pi}{\lambda z}(x\xi+y\eta)}dxdy \tag{3.11}$$

which further can be written in Fourier Transform form as,

$$A_I(\xi,\eta) = \int_{-\infty}^{+\infty}\int_{-\infty}^{+\infty} A_o(x,y)e^{\frac{i2\pi}{\lambda z}(x\xi+y\eta)}dxdy \tag{3.12}$$

The holograms synthesized according to Eq. 3.12 by means of Fourier transformation is called as synthesized Fourier holograms . Fourier and Fresnel holograms can be computed through Discrete Fourier transforms. Schemes of using computer

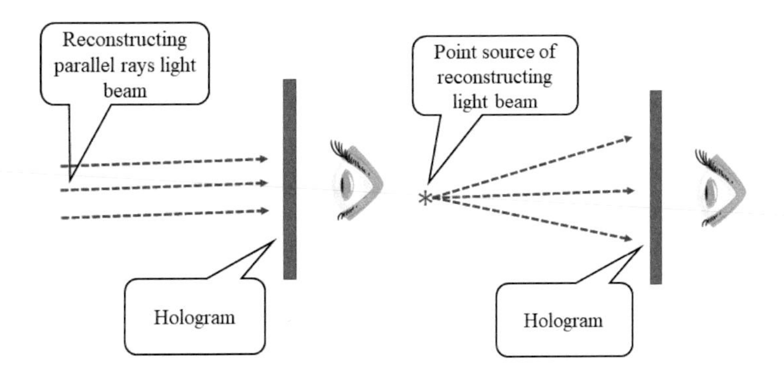

Figure 3.3 Schematic geometry for computer generated display of a) Fourier hologram and b) Fresnel hologram visualization.

generated display Fresnel and Fourier holograms for visual observation of virtual objects are shown in Fig. 3.3 a) and b), respectively. Fresnel holograms differ from Fourier holograms when comes to object wave front reconstruction, as Fresnel hologram has focusing properties and are capable of reproducing the finite distance from the observation plane to the object plane. On the other hand, reconstruction of Fourier holograms can be done only by spherical reconstruction beam and reconstructed image is observed in the plane of the reconstructing beam point source.

Mathematically, reconstructing an object using Fresnel and Fourier holograms is described by the inverse Fresnel and Fourier transformations process respectively. When holograms are observed visually, these transformations are performed by the optical system of observer's eye. In computer generated hologram the shifting of 3D to 2D process fails to take into account the effect of object depth relief on the wave front. Fresnel holograms also use only a single distance value from the object to the observation plane rather than object relief depth. Nevertheless, it is still possible to synthesize holograms, which are capable of reconstructing 3-D images for visual observation and retain the most important property of holographic visualization that is the depth information of the object using the composite computer generated stereo and macro-holograms and programmed diffuser holograms. The stereo holograms are holograms synthesized for viewing by left and right eye and to reconstruct the corresponding stereo views of the object. The composite stereo holograms are composed of multiple holograms that reproduce different aspects of the object as observed from different observation positions. Macro-holograms are large composite stereo-holograms that cover wide observation angle which can be used as a wide observation window. The diffuser holograms are computer generated Fourier holograms which imitate properties of diffuse surfaces to scatter irradiation non-uniformly in space and provide object surface shape.

3.3 REALIZATION OF COMPUTER-GENERATED HOLOGRAPHY

The computer generated holography can be realized using three different techniques. The first one is *Amplitude only media*, the next one is *Phase only media* and the third technique is *Combined Amplitude and phase media* respectively. i) In amplitude-only media, the light intensity transmission or reflection factor is the controll parameter . This is the most common and available technique, where the conventional silver halide photographic emulsion(Agfa Gaevert 10E 75 or 8e 75) is used in photography and optical holography .ii) In the second case of phase-only media, the optical thickness of the medium can be controllable by either varying the refractive index medium or the physical thickness, or both of them. The phase-only media for recording CGH(Computer Generated Hologram) include thermoplastic materials, photo resists, bleached photographic materials, photo polymers, etc. Recently, micro-lens array and micro-mirror array technology has emerged for recording computer generated holograms. The third and another important technique is the combined media (phase and amplitude)where it allows independent control of both the light intensity transmission factor and the optical thickness. At present, these recording techniques use photographic materials with two or more layers sensitive to radiation of different wavelengths which permits the user to control the transparency of certain layers and the optical thickness of by exposing each layer to its wavelength independently. But, for the mathematical hologram, digitally controlled hologram recording devices are required for modulating optical parameters of these media. Unfortunately, no such special purpose devices are available, instead the computer printer/plotters, display devices and other improvised means such as photo- and e-beam (electron) lithography are being used. These devices perform only binary or two level modulation of the medium's optical parameters. In case of amplitude- only and phase-only media the controlled optical parameters may assume only two values which can be then referred to as binary media mode. In case of binary mode, the amplitude and phase media will be inefficient in terms of the medium information capacity. This is due to the fact that writing information on them is defined only by their spatial degrees of freedom (their resolution power), whereas in case of combined amplitude and phase media,the degrees of freedom related to a transmission (reflection, refraction) factor can also be used. The major advantage of binary mode is the simplicity of media exposure and copying recorded holograms. For achieving a CGH by modulating phase only and phase and amplitude one can use Spatial Light Modulators(SLM) in conjunction with a computer. In following section the demonstration of a Computer Generated Hologram is explained using a Spatial Light Modulator.

3.3.1 REALIZATION OF COMPUTER GENERATED HOLOGRAM USING A SPATIAL LIGHT MODULATOR

Computer generated Hologram (CGH) is digitally generated with objects as letter H and C as shown in Fig. 3.4(a) and 3.4(b), where the letter $H$is kept as in plane and C is kept at depth of 5mm. With these two displaced letters as an object, a hologram is digitally generated with tilted reference plane wave as shown in Fig. 3.4(c). Consider

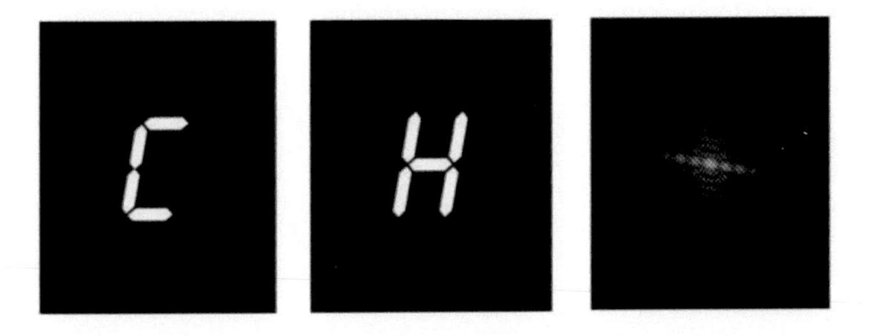

Figure 3.4 The letters C and H for generating computer generated hologram

the experimental geometry shown in Fig 3.5 where an un-expanded laser beam passes through a halfwave plate to produce an x-polarized beam. Now, the CGH shown in Fig. 3.5 is displayed on Spatial Light Modulator (SLM). Since the SLM used in the experiment is sensitive to x-polarized light, the light from the laser source (632.8 nm, Y-polarized) is converted to x-polarized with help of a half wave plate (HWP, 22.5 degree) and a linear polarizer (LP1). After Spatial filtering and collimation with Lens $L(f = 200mm)$, the beam becomes normal incident to the plane of SLM. The Lens L shown in Fig. 3.5 does the Fourier Transform of the hologram displayed on SLM at CCD plane which is at a distance f from the Lens. A polarizer at the output port is used to maintain the contrast of the reconstructed image, and an aperture is also used to select the first order of the diffracted pattern from SLM for filtering out the zeroth and higher orders. The results of object reconstruction from a CGH are shown

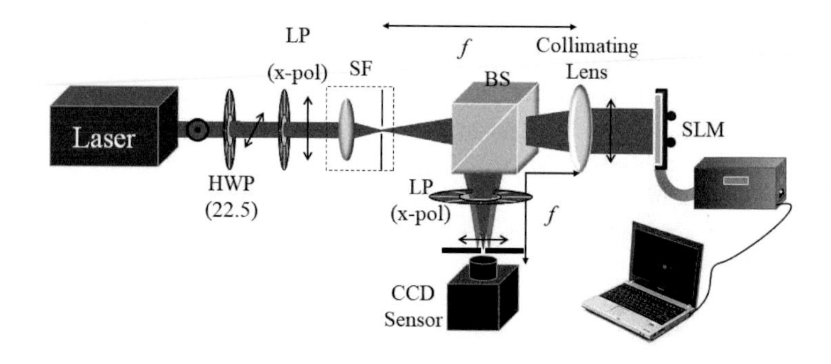

Figure 3.5 Schematic geometry for recording computer generated hologram using a spatial light modulator.

in Fig. 3.6. As mentioned earlier the letter C is at depth of $5mm$ off plane. The Fig 3.6(a) shows the reconstructed letter $Catdz = -5mm$. Fig. 3.6(b) shows the result of reconstruction of the in plane $(dz = 0)$ letter H. In this we can see that the letter H and its conjugate are in focus. The Fig. 3.6(c) shows the results at $dz = 5mm$ when conjugate of letter C is in focus. A more simple experiment for demonstration of a computer generated hologram even in undergraduate college environment is clearly demonstrated by Vijayakumar et.al[22].

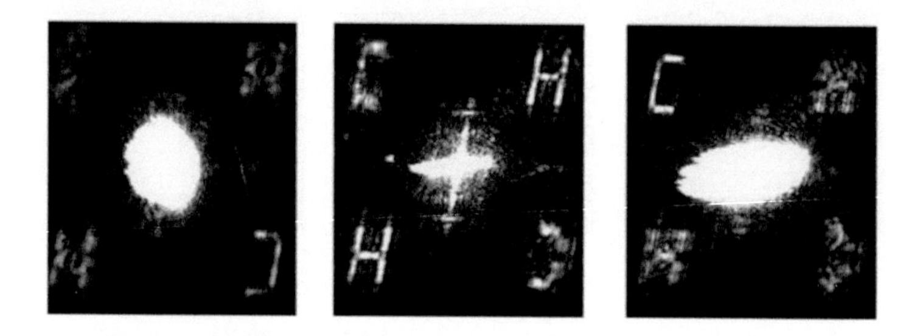

Figure 3.6 Results obtained from computer generated hologram.

4 Photorefractive dynamic holography

4.1 INTRODUCTION

The photorefractive dynamic holography uses a different type of non-linear optical effect known as photorefractive effect. The photorefractive effect occurs in certain crystals of sillienite family like Bismuth Germanium Oxide ($Bi_{12}GeO_{20}$ or BGO), Bismuth Silicon Oxide ($Bi_{12}SiO_{20}$ or BSO), Bismuth Titanium Oxide ($Bi_{12}TiO_{20}$ or BTO) and polar crystals like Barium Titanate (BaTiO3), Lithium Niobate (LiNbO3) etc[23-26]. The photorefractive effect occurs in these crystals when the refractive indices change due to optically induced spatial distribution of electrons and holes when two beams interfere inside these crystals. The photorefractive effect phenomena is different from conventional non-linear effects like phase conjugation, optical limiting, second harmonic generation as these non-linear effects are due to intensity of laser beams, where as PR effect is due to realtive intensities of light beam(Interference pattern). Thus the photorefractive effect can not be described using non-linear susceptibility χ^n for any values of n and the advantage is that even with 2mW power laser can give photorefractive effect and high intense laser beams are not necessary. Since the photorefractive effect requires realtive intensities inside the photorefractive crystal two wave mixing is a convenient method in which one beam can be signal or object beam and other one probe or reference beam similar to a holography[27-30]. In such two wave mixing geometry, the photorefarctive crystal becomes recording medium with one major deviation from conventional holographic recording medium that is dynamic recording and continuous reconstruction without the need for wet processing. Photorefractive effect based dynamic holography is real-time and recording and reading the holograms are done simultaneously without the need of reconstruction. The two wave mixing phenomena inside a photorefractive crystal can be use to record real-time dynamic holograms(Random Access Memory Type), for image amplifications, phase conjugation, opto-electronic correlators etc, with relatively low intensity requirements via the photorefractive effect. In particular, photorefractive crystals (PRCs) are widely used for real-time holographic recording because of their practically unlimited recyclability and they can be recorded and erased in this media with sufficiently high sensitivity. The PR crystals like BSO/BTO/BGO possess high sensitivity for volume holographic grating formation that enables them to record holograms in the visual region of the spectrum with continuous wave lasers like He-Ne, He-Cd, Argon Ion etc. Also, these crystals are available in large sizes and good quality. There were many innovative geometries have been proposed by various authors to record holograms in these crystals with and without externally applied electric field across the crystal in both two and four wave mixing geometries respectively. In some cases, electric field is applied across the photorefractive crystal for increasing the diffraction efficiency of dynamic hologram.There are many books and research papers on photorefractive effect and its applications[23-30] and in this chapter we describe the photorefractive dynamic holography using crystals of sillienite family such as Bismuth Silicon Oxide($Bi_{12}SiO_{20}$) and Bismuth Titanium Oxide ($Bi_{12}TiO_{20}$).

4.2 PRINCIPLE OF DYNAMIC PHOTOREFRACTIVE HOLOGRAPHY

4.2.1 PHOTOREFRACTIVE EFFECT

The phenomena of photorefractive effect is illustrated schematically in Fig. 4.1. We assume that a photorefractive crystal is illuminated by two beams of same frequency (In dynamic holography, these would be the signal and reference beams)and they interfere in the crystal to produce spatially modulated intensity distribution $I(x)$. This interference pattern between the object and reference beams results in a pattern of dark and light fringes throughout the crystal (Fig. 4.1 a). Now, the free charges(either electrons or holes) are generated through photoionization at a rate that is proportional to local value of intensity. The electrons or holes present, in regions of photorefractive crystal where a bright fringe is present, absorb the light and get photoexcited from an impurity level into the conduction band of the photorefractive crystal, leaving net positive or negative charges. The impurity levels have an energy intermediate between the energies of the valence band and conduction band of the PR crystal and so, the electrons or holes in conduction band of crystal, diffuse throughout the crystal. Since the charges are excited in the region of bright fringes, the net charge diffusion current is towards the dark-fringe regions of the PR crystal and in the conduction band, the charges(electrons and holes), recombine and return to the impurity levels. The rate at which this recombination takes place determines how far the charges(electrons/holes) diffuse, which in turn determines strength of the photorefractive effect in that crystal. Now in the impurity level, the electrons/holes are trapped and can no longer move unless re-excited back into the conduction band by light. This spatially varying space charge distribution $\rho(x)$ will in turn produce space charge field (E_{sc}) in the crystal as shown in Fig. 4.1c. It can be seen that the maxima of space charge field is shifted by 90^0 with respect to charge density distribution$\rho(x)$. This shift is due to Maxwell's equation $\nabla.\mathbf{D} = 4\pi\rho$ and for this situation it will be $\frac{dE_{sc}(x)}{dx} = \frac{4\pi\rho(x)}{\varepsilon}$. Since the electrons and holes are trapped and immobile, the space charge field (E_{sc}) remains even after the removal of interference patterns. This internal space charge field (E_{sc}), due to electro-optics effect(Pockel's effect) causes the refractive index of the crystal to change in the regions where the field is strongest and this causes a spatially varying refractive index grating Δn shown in Fig.1d, to occur throughout the crystal. The refractive index change Δn is shifted by 90^0 with respect to the intensity distribution $I(x)$. This dynamic phase shift is responsible for transfer of energy from one beam to another. This dynamic grating can diffract, light shone into the crystal, and the resulting diffraction pattern recreates the original pattern of light stored in the crystal which is either the object beam/reference

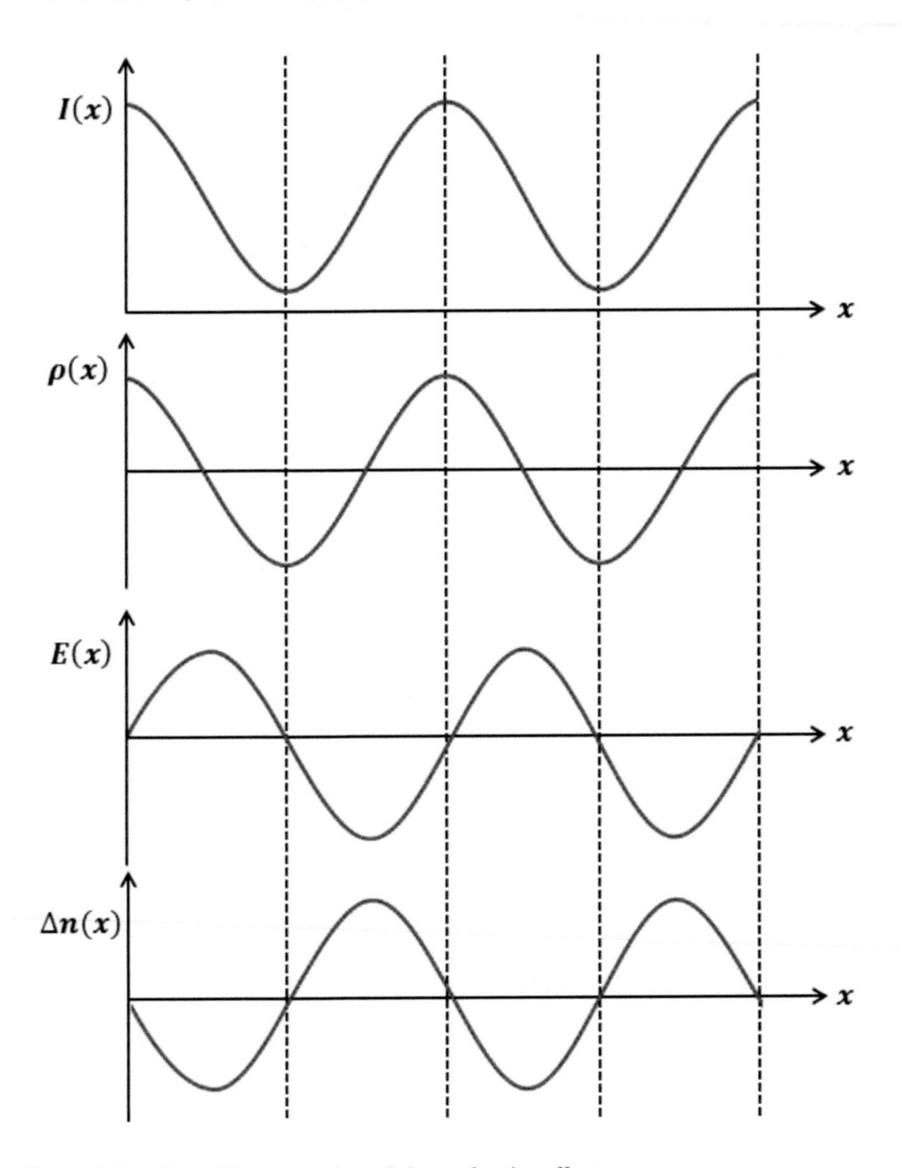

Figure 4.1 a, b, c, d Demonstration of photorefractive effect

beam. Thus a photorefractive crystal can act as dynamic hologram if both object and reference beam interfere in the photorefractive crystal.

4.2.2 THEORETICAL EXPLANATION

The theoretical explanation of photorefractive effect can be explained by Kukhtarev et.al's model[24], where they consider only one type charge that is electron.

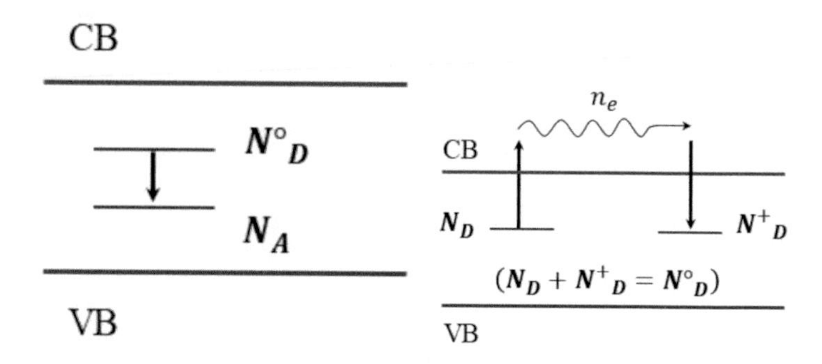

Figure 4.2 Explanation of photorefractve effect using band diagram 4.2a, 4.2b

Consider, Fig. 4.2a where, the photorefractive crystal contains N_A acceptors and N_D^0 donors per unit volume and assume that the acceptors level are completely filled with electrons. These electrons are actually fallen from donor levels and these filled up acceptor levels cannot be ionized by either thermally or by optical effects. this ensures that at temperature level $T = 0$, in absence of any optical field each unit volume of crystal contains N_A ionized donors, N_A electrons bound to acceptor impurities, and $(N_D^0 - N_A)$ neutral donor levels which can participate in photorefractive effect. We further assume that the electrons can be excited into conduction band from donor level using optical or thermal effects which is as shown in Fig. 4.2 b. Here, n_e(number of densities in conduction band), N_D^+ and N_D denote the ionized and un-ionized donors respectively. Also, it is important to note that $(N_D + N_D^+)$ must be equal to N_D^0 and N_D^+ is not equal to n_e, because some donors move from donors to acceptors due to migration within photorefractive crystal which results in electrically neutral region. The variations in population levels can be described using rate equations,

$$\frac{\partial N_D^+}{\partial t} = (sI + \beta)(N_D^0 - N_D^+) - \gamma n_e N_D^+ \tag{4.1}$$

$$\frac{\partial n_e}{\partial t} = \frac{\partial N_D^+}{\partial t} + \frac{1}{e}(\nabla.\mathbf{j}) \tag{4.2}$$

where s is a constant proportional to the photoionization cross section of a donor, β is the thermal generation rate, γ is recombination co-efficient, $-e$ is the electronic charge, and j is electrical current density. The Eqn.4.1 shows that the ionized donor concentration can increase by thermal ionization or by photoionization of un-ionized donors and can decrease by recombination. Also, Eqn.4.2 describes that the mobile electron concentration can increase in small region because of the ionization of donor atoms or the flow of electrons in to local region. The flow of current can be described by following Equation as,

$$\mathbf{j} = n_e e \mu \mathbf{E} + eD\nabla n_e + \mathbf{j}_{ph} \tag{4.3}$$

where μ is the electron mobility, $D = k_B T \frac{\mu}{e}$ is diffusion co-efficient, $\mathbf{j}_{ph}$ is photovoltaic contribution to the current. The photovoltaic contribution results from tendency of photoionization process to eject the electron in a preferred direction in an anisotropic crystal. The electric field component $\mathbf{E}$ appearing in Eqn.4.3 is the static electric field appearing within the crystal due to any applied voltage or to any charge separation within the crystal and it must staify Maxwell's equation as,

$$\varepsilon_{dc} \nabla . \mathbf{E} = -4\pi e(n_e + N_A - N_D^+) \tag{4.4}$$

The term ε_{dc} represents static dielectric constant of the crystal. Then the modification of optical properties is described in terms of change in optical frequency dielectric constant by an amount,

$$\Delta\varepsilon = \varepsilon^2 r_{eff} |E| \tag{4.5}$$

The optical field $\vec{E}_{opt}$ obeys the wave equation given by,

$$\nabla^2 \vec{E} + \frac{1}{c^2} \frac{\partial^2}{\partial t^2} (\varepsilon + \Delta\varepsilon) \vec{E}_{opt} = 0 \tag{4.6}$$

Equations from 4.1 through to 4.6 describe basic photorefractive equations.

4.2.3 TWO WAVE MIXING IN PHOTOREFRACTIVE CRYSTALS

The two wave mixing in a photorefractive crystal/material can be considered, similar to two beams(Object and Reference) intereference in holography. The holographic plate is the recording medium in holographic recording geometry and in dynamic photorefractive holography, the photorefractive crystal is the recording medium. In photorefractive holography the signal beam represents object beam and pump beam represents reference beam. The mixing(in other words interference) of signal(object) and pump(reference) beams inside the photorefractive crystal is known as two beam coupling. This is because the two beam coupling process can be used to amplify a weak signal beam carrying information about an object(image) using the pump beam at the rate of exponential gain of 10 per centimeter of crystal size. Consider Fig. 4.3 in which a typical two beam coupling geometry is shown. It can be easily recognized that the geometry is similar to conventional holography where the recording holographic plate is replaced with a PR crystal. Let $A_s(z)e^{i\hat{k_s}.\hat{r}}$, $A_p(z)e^{i\hat{k_p}.\hat{r}}$, represent signal(object) and pump(reference) beams respectively and they interfere inside the crystal to form non-uniform intensity pattern(bright and dark fringes). Fig. 4.1 a shows such pattern and because of that a refractive index grating is formed inside the crystal. But this grating is displaced from the intensity distribution in the direction of positive or negative depending upon the dominant charge carriers and sign of electro-optic coefficient. Because of this phase shift the light scattered from pump beam(A_p) interfers constructively with signal beam(A_s) and light scattered from signal beam interfers destructively with pump beam there by amplifying signal beam and attenuating pump beam. This process can be mathematically described and the total field within the photorefractive crystal is,

$$E_{opt}(\mathbf{r}, t) = [A_p(z)e^{i\hat{k_p}.\hat{r}} + A_s(z)e^{i\hat{k_s}.\hat{r}}]e^{-i\omega t} + c.c \tag{4.7}$$

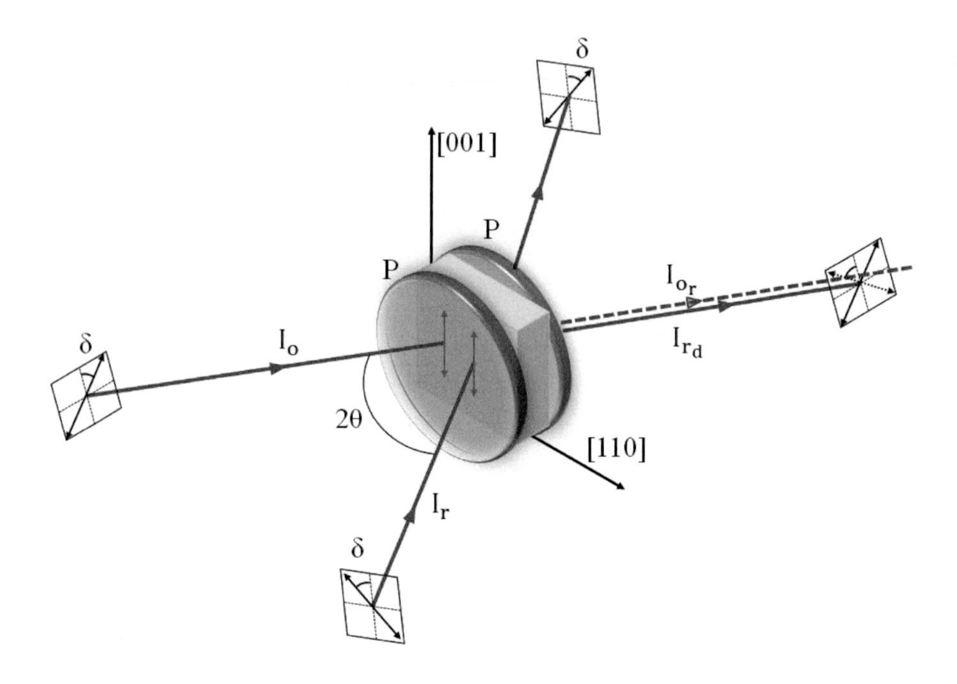

Figure 4.3 Two wave mixing in BSO(Bismuth Silicon Oxide) crystal.

where, (A_p),(A_s) are slowly varying amplitude functions of pump and signal beams respectively. The total intensity inside the crystal is given by,

$$I = (\frac{n_0 c}{4\pi})[(|A_p|^2 + |A_s^2) + (A_p A_s^*)(\hat{e_p}.\hat{e_s})e^{ik_g x} + c.c)] \tag{4.8}$$

In above equation $\hat{k_g} = k_g \hat{x} = (\hat{k_p} - \hat{k_s})$ with $\hat{e_p}, \hat{e_s}$ representing polarization unit vectors of pump and signal beams respectively and they are assumed to be linearly polarized. The $\hat{k_g}$ is the grating vector. The intensity distribution also can be written as,

$$I = I_0[1 + m cos(k_g x + \phi)] \tag{4.9}$$

with $m = \frac{2|I_1|}{I_0}$ known as modulation index and $\phi = tan^{-1}(\frac{ImI_1}{ReI_1})$. To determine the changes in optical properties of photorefractive materials due to the presence of pump and signal fields, the equations from 4.1 through to Eqn.4.4 have to be solved for finding static electric field **E** induced by the relative intensity distribution of signal and pump beams given by Eqn.4.8. With this static electric field the change in optical frequency dielectric constant can be calculated. But, the equations to be solved that is from 4.1 through to 4.4 are nonlinear for solving that consider following expressions as,

$$E = E_0 + (E_1 e^{ik_g x} + c.c) \tag{4.10}$$

$$j = j_0 + (j_1 e^{ik_g x} + c.c) \tag{4.11}$$

$$n_e = n_{e0} + (n_{e1} e^{ik_g x} + c.c) \tag{4.12}$$

$$N_D^+ = N_{D0}^+ + (N_{D1} d^{ik_g x} + c.c) \tag{4.13}$$

In above expressions, $\vec{E} = E\hat{x}$ and $\vec{j} = j\hat{x}$ and have assumed the quantitites such as E_1, j_1, n_{e1} and N_{D1} are small as product of any two of them can be neglected. Now, substituting equations from 4.10 through to 4.13 in respective equations from 4.1 through to 4.4 and equating the terms with common x depenedences, we get following expressions,

$$(sI_0 + \beta)(N_D^0 - N_{D0}^+) = \gamma n_{e0} N_{D0}^+, \tag{4.14}$$

$$j_0 = Constant \tag{4.15}$$

$$j_0 = n_{e0} e \mu E_0 + j_{ph} \tag{4.16}$$

$$N_{D0}^+ = n_{e0} + N_A \tag{4.17}$$

The values of mean electron density n_{e0} and mean ionosed donor density N_{D0}^+, can be obtained by solving Eqns.4.14 and 4.17 as,

$$n_{e0} = \frac{sI_0 + \beta)(N_D^0 - N_A)}{\gamma N_A} \tag{4.18}$$

$$N_{D0}^+ = N_A \tag{4.19}$$

Also, $n_{e0} << N_A$ has to be satisfied. The other two equation i.e 4.15 and 4.16 determine the mean current density j_0 and mean field E_0. Consider that the photovoltaic contribution j_{ph} is negligible for the photorefractive crystal then, mean electric field E_0 will depend upon the applied electric field connected across it and if there is no applied field then both E_0 and j_0 vanish. To find the static electric field E, we consider restructured equations 4.1 and 4.2 as,

$$\frac{\partial n_e}{\partial t} = (sI + \beta)(N_D^0 - N_D^+) - \gamma n_e N_D^+ + \frac{1}{e}(\frac{\partial \mathbf{j}}{\partial x}) \tag{4.20}$$

Re organizing Eqn.4.3 to find $\frac{\partial \mathbf{j}}{\partial x}$ we get,

$$\frac{\partial \mathbf{j}}{\partial x} = n_e e \mu_e \frac{\partial \mathbf{E}}{\partial x} + eD \frac{\partial^2 n_e}{\partial x^2} \tag{4.21}$$

Now, we restructure above equation as,

$$\frac{\partial \mathbf{E}}{\partial x} = \frac{\frac{\partial \mathbf{j}}{\partial x}}{n_e e \mu} - \frac{eD}{n_e e \mu_e} \frac{\partial^2 n_e}{\partial x^2} \tag{4.22}$$

Integrating Eqn.4.22 we get the value of electric field component,

$$E = \frac{j - eD\frac{\partial n_e}{\partial x}}{n_e e \mu_e} \tag{4.23}$$

we also know that $n_e << N_A, (N_D - N_A), N_D^+ = N_A$ in the steady state and the equation for number of free electron becomes,

$$n_e = N_0[1 + mcos(\vec{k}_g.\vec{x})] \tag{4.24}$$

where, $N_0 = g(I_0)\tau_R$, $g(I_0)$ representing linear generation rate and τ_R is the free electron life time, D is the diffusion co-efficient and by Einstein's relation it is equal to $\frac{k_B T \mu}{e}$. Substituting respective values for n_e and simplifying Eqn.4.23 becomes,

$$E = \frac{J}{e\mu_e}\frac{1}{(1 + mcosk_g x)} - [\frac{Dk_g}{\mu_e}]\frac{msin(k_g x)}{(1 + mcosk_g x)} \tag{4.25}$$

Equation 4.25 represents electric field component.

Normally one can apply electric field across the crystal to enhance the efficiency of two wave mixing or in certain photorefractive crystals of sillienite family like $Bi_{12}SiO_{20}$(Bismuth Silicon Oxide) and $Bi_{12}TiO_{20}$ (Bismuth Titanium Oxide) the two wave mixing can be achieved without applied electric field. If there is an applied voltage V across the crystal length L, then the electric field component in equation 4.25 becomes,

$$\frac{1}{L}\int_0^L Edx = \frac{j}{e\mu_e}\frac{1}{L}\int_0^L \frac{1}{(1 + mcosk_g x)}dx - [\frac{Dk_g}{\mu_e}]\frac{1}{L}\int_0^L \frac{msin(k_g x)}{(1 + mcosk_g x)}dx \tag{4.26}$$

where the integral values for large number of fringes will give values,

$$\frac{1}{L}\int_0^L \frac{1}{(1 + mcosk_g x)}dx = \frac{1}{\sqrt{(1 - m^2)}} \tag{4.27}$$

and,

$$\frac{1}{L}\int_0^L \frac{msin(k_g x)}{(1 + mcosk_g x)}dx = 0 \tag{4.28}$$

substituing Equations 4.27, 4.28 in equation 4.26 the value of flow of current j becomes

$$j = \sqrt{(1 - m^2)}e\mu_e N_0 E_A \tag{4.29}$$

where $E_A = \frac{V}{L}$ is the applied electric field. It may be noted here that for applied electric field, the conductivity of cosinusoidal illumination is reduced by a factor $\sqrt(1 - m^2)$ related to conductivity at the same average intensity. Applying j values in Eqn.4.23 and simplifying we get,

$$E_{sc} = E_A\frac{\sqrt{(1 - m^2)}}{(1 + mcosk_g x)} - E_D\frac{msin(k_g x)}{(1 + mcosk_g x)} \tag{4.30}$$

THe value of $E_D = (\frac{k_B T}{e})k_g$ is called characteristic field and in equation 4.30 the electric field E_{sc} replaces E, known as space charge field. If there is no applied electric field i.e $E_A = 0$ then the space charge field E_{sc} reduces to diffusion field as charge migration takes place only due to diffusion in the photorefractive crystal whenever two beams mix in the crystal. This space charge field created due to charge separation in photorefractive crystal changes the refractive index value Δn via the electro-optic Pockel's effect and is equal to,

$$\Delta n = \frac{1}{2} n_0^3 r_{eff} E_{sc} \tag{4.31}$$

with r_{eff} is the effective electro-optic co-efficient and n_0 represents the unperturbed reractive index. This refractive index change results in live diffraction grating with an efficiency equalent to,

$$\eta = e^{-(\frac{\alpha t}{cos\theta t_B})sin^2(\frac{\pi t}{\lambda cos\theta t_B}\Delta n)} \tag{4.32}$$

where α is the absorpion co-efficient of the photorefractive crystal, t is the thickness of crystal, λ is the wavelength of light used and θ_B is the Bragg angle inside the crystal. Since in photorefractive effect as mentioned earlier the weak signal(Object) beam is amplified by pump(reference) beam due to constructive interference between these two in the crystal the efficiency of two wave mixing or dynamic hologram is given by,

$$\Gamma = \frac{2\pi}{\lambda} n_0^3 r_{eff} E_{sc} \tag{4.33}$$

4.3 EXPERIMENTAL TECHNIQUES OF PHOTOREFRACTIVE DYNAMIC HOLOGRAPHY

4.3.1 INTRODUCTION

Conventional holographic recordings are carried out using time delayed construction and reconstruction process using wet processing method. In case of dynamic photorefractive holography, the holographic images are readout and recorded simultaneously using self-diffraction process in a two wave mixing geometry. Though there are techniques for recording dynamic holograms from photorefractive crystals, but most of them use the self-diffraction phenomena. In this scheme, the readout and recording of holograms occur simultaneously and the reference beam itself acts as continuous readout beam at the same time acting as recording beams along with object beam. The self- diffraction phenomena gives rise to to energy transfer between the recording beams emerging from crystal during holographic recording. As mentioned in previous chapter in two wave mixing in the photorefractive crystals the signal or object beam gets amplified when the pump or reference beam constructively interfere and the pump or reference beam becomes weaker as they destructively interfere in the crystal. Thus when two beams interfere in the photorefractive crystals they create a dynamic grating due to which the self-diffraction phenomena takes place.

Now the amplified diffracted object or signal beam and depleted pump or reference beam simultaneously propagate along with direct object or signal and pump or reference beams. Due to this the efficiency or contrast of diffracted hologram becomes low. In cubic paraelectric crystals of the sillenite family like $Bi_{12}SiO_{20}$(Bismuth Silicon Oxide), $Bi_{12}TiO_{20}$(Bismuth Titanium Oxide), and $Bi_{12}GeO_{20}$(Bismuth Germanium Oxide),in addition to amplification of signal/object beams, polarization rotation of diffracted beams with respect to direct beams without change in energy can occur and this phenomena is called an-isotropic self diffraction. The anisotropic self-diffraction phenomena(Appendix C) is the most convenient process for recording photorefractive dynamic holography and it occurs due to certain orientation of photorefractive crystal. The an-isotropic self-diffraction process solves an important technical difficulty in recording dynamic holograms using photorefractive crystals. In photorefractive media due to self-diffraction process both readout and recording of holograms occur simultaneously where only energy of signal/object beam is amplified. But to get diffracted hologram, the reconstructing holographic beam or readout beam, must satisfy the Bragg condition and that becomes stringent due to small sizes of these photorefractive crystals allowing small diffraction angles. Also, simultaneous hologram readout and recording requires same wavelength at the same time the angle of incidence of hologram readout beam must coincide with angle of reference beam at the crystal plane. These stringent conditions make the separation of direction of diffracted beams with respect to direct beams more complex. One can solve this in two ways, i) using nonlinear holographic recording utilizing wide band reconstruction process with other wavelengths but, this process has low sensitivity and diffraction efficiency and ii) using phase conjugation method which requires 4 beams i.e two beams for recording a hologram and one beam for hologram recontruction and other diffracted beam respectively. So these two processes will not be practical and the most efficient method could be the an-isotropic self diffraction phenomena occuring in certain cubic paraelectric crystals of the sillenite family like $Bi_{12}SiO_{20}(BSO)$(Bismuth Silicon Oxide), $Bi_{12}TiO_{20}(BTO)$(Bismuth Titanium Oxide), and $Bi_{12}GeO_{20}(BGO)$(Bismuth Germanium Oxide). Fig. 4.3 shows a typical two wave mixing geometry in a BSO crystal where anisotropic self diffraction phenomena occurs. In that the polarization of diffracted beams are rotated with respect to direct beams and is demonstrated in Fig. 4.3. The anisotropic self-diffraction phenomena occurs due to crystallographic orientation and the most important advantage of anisotropic self diffraction phenomena is the separation of diffracted beams with respect to direct beams in terms of their polarization orientation(Appendix C). This will immensely enhance the diffraction efficiency of dynamic holograms.

4.3.2 PHOTOREFRACTIVE DYNAMIC HOLOGRAPHY USING $Bi_{12}SiO_{20}$ (BISMUTH SILICON OXIDE)

The practical dynamic hologram recording geometry using anisotropic self-diffraction phenomena in BTO was proposed by A A Kamshalin et.al[29], for measuring vibration of a speaker using time average holographic interferometry. This geometry does not require application of electric field across the photorefractive crystal as the anisotropic self diffraction occurs due to diffusion field. Later, Troth and Dainty[30] modified the geometry(Fig. 4.3) to implement time average holographic interferometry using BSO(Bismuth Silicon Oxide). In both geometries the photorefractive crystal is sandwiched between two polrizers as shown in Fig. 4.3. We follow Troth and Dainty's[30] dynamic holography model using Bismuth Silicon Oxide as photorefractive holography medium in this section. The anisotropic

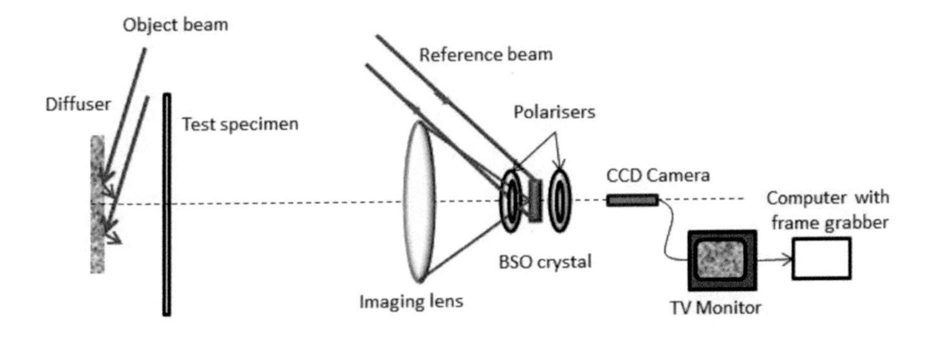

Figure 4.4 Experimental geometry to record a dynamic photorefractive hologram using BSO crystal

self-diffraction(Appendix C) produces a rotation in linear polarization of diffracted beams with respect to direct beams which helps us to separate the direct and diffracted beams by rotating the analyzer at the output. The experimental geometry shown in Fig. 4.4 is similar to one developed by Troth and Dainty[30]. Fig. 4.4 shows a light beam from a green laser ($\lambda = 532nm$) and falls on a diffuser screen. The scattered speckle beam from the diffuser passes through a plastic scale (Transparent object) made up of PMMA(Poly Methyl Meta Crylate)which is a photoelastic material and then, imaged on to the BSO crystal sandwiched between two polarizers. The image in the BSO crystal is captured by a CCD camera which is shown in Fig. 4.4. The reference beam also falls directly on to the BSO crystal through same polarizer like the object beam so as to keep both object and reference beams in same state of polarization before they interefere in the crystal. Due to two wave mixing phenomena, and anisotropic self diffraction the hologram of object can be clearlty seen without the disturbance of direct beams like in conventional in-line holography as well as off-axis holography where the real and virtual images superpose. It may be noted here that even in digital holography where we will be discussing in 5th chapter suppression of direct beams need proper technique. Fig. 4.5 shows a typical photorefractive dynamic hologram of a portion of transparent scale without any disturbances of direct or conjugate beams due to ani-isotropic self diffraction. Thus holographic images obtained using dynamic photorefractive holography are free from superposing direct beams and the main draw back is the avaialability of high quality photorefractive crystals and need to downsize obhjects to 1/4th of

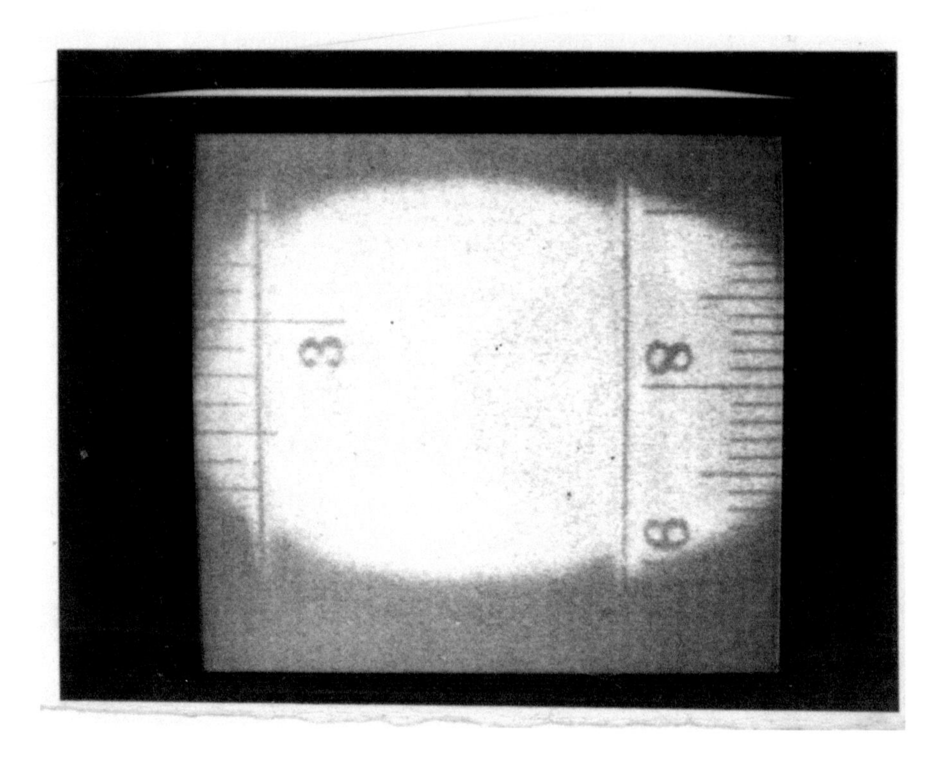

Figure 4.5 Dynamic photorefractive hologram of a portion of transparent plastic scale made up of photoelastic material PMMA

original object. Fig. 4.4 is a two in one photorefractive holographic geometry where both a phase object(transparent scale) and 3D object(position of diffuse screen) can be recorded separately. In the market photorefractive holographic cameras with BSO crystal as recording medium are available and are being used widely in the industrial environment.

5 Digital holography

5.1 INTRODUCTION

In computer generated holography, we numerically construct hologram of an object and then physically reconstruct the hologram with a laser beam. On the other hand digital holography records physically the hologram of an object on a CCD/CMOS array and the holographic image is reconstructed numerically using a computer. Thus Computer Generated Holography(CGH) and Digital holography techniques are complimentary. In the early days of holography, some research was done using television transmission of holograms. In that the holograms were performed directly on the photosensitive layer of a TV camera tube, and the interference signal was transmitted through the television channel. The reconstruction of hologram was written on a transparent surface at the receiving station and the transmitted hologram was reconstructed optically. In this way, hologram was recorded optoelectronically and reconstructed optically. There are also some works realted to reconstructing objects from holograms using mathematical methods. In these researches, an optically enlarged part of a conventionally recorded in-line-hologram is used for reconstruction. There are many researchers who contributed to the development of digital holography[31-35] and Thomas Kries[36] book covers most of basic principles and applications of digital holography. In this chapter we follow similar style of Thomas Kries[36] especially the construction and reconstruction processes of digital holography.

5.2 PRINCIPLE OF DIGITAL HOLOGRAPHY

The expermental geometry to record digital holography is shown in Fig. 5.1 in which a diffusely reflecting object is illuminated by part of an expanded laser beam(He-Ne or Nd: Yag) from beamsplitter. The other part of expanded laser beam and the scattered object beam from the difuse object now interferes on CCD array as shown in Fig. 5.1. In conventional holography, a plane reference wave and the diffusely reflected object wave interfere at the surface of a photosensitive medium(Agfa Gaevert 10 E 75 film) but in digital holography object and reference beam are recorded as interference pattern in CCD/CMOS array. The object beam in conventional hologram is reconstructed by illumination of the processed hologram with the same reference wave which was used to record it. Then the real image can be seen if a screen is placed at a distance d behind the hologram. On the other hand, in case of digital holography, by using numerical reconstruction process, the real image can be reconstructed. The amplitude and phase distribu- tion in the plane of the real image can be described using Fresnel-Kirchhoff integral. Considering that a plane wave illuminates the hologram located in the plane $z = 0$, with an amplitude transmittance T(x, y), the Fresnel-Kirch- hoff' integral results in the complex amplitude $A(\xi, \eta)$ which

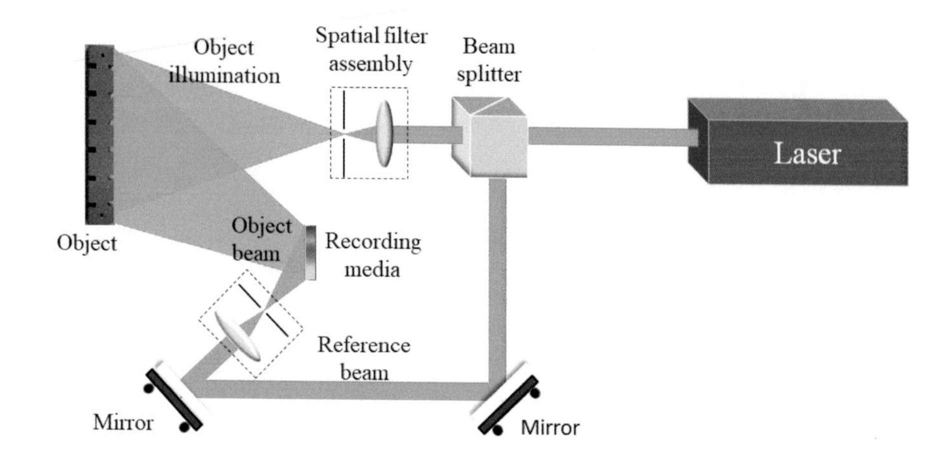

Figure 5.1 General experimental geometry for recording an Off-axis digital hologram

is given by,

$$A(\xi,\eta) = \frac{ia}{\lambda d} e^{\frac{-i\pi}{\lambda d}(\xi^2+\eta^2)}$$
$$\int\int_{x,y} e^{\frac{-i\pi}{\lambda d}(x^2+y^2)} T(x,y) e^{\frac{i2\pi}{\lambda d}(x\xi+y\eta)} dxdy \tag{5.1}$$

where "a" is the amplitude of the incident wave and Eq.5.1 is Fresnel approximation formula and is valid if the following condition is fulfilled,

$$d^3 >> \frac{\pi}{4\lambda}[(x-\xi)^2+(y-\eta)^2]^2 \tag{5.2}$$

In Eqn.5.2, for the inequality the maximum possible value of $(\xi - x)^2$ and $(\eta - y)^2$ must be considered. For a wavelength $\lambda = 600nm$ and typical hologram dimensions of $(\xi - x)$ max $= (\eta - y)$ max $= 0.5cm$, d must be larger than 15 cm. The intensity of real image is $I_H(\xi,\eta) = |A(\xi,\eta)|^2$ which can be reconstructed in digital holography by digitizing function $A(\xi,\eta)$ with its amplitude transmission function $T(x,y)$ is sampled on a rectangular raster of $N \times N$ matrix points, with steps Δx and Δy along the coordinates. Replacing the terms ξ and η by $r\Delta\xi$, and $s\Delta\eta$, where r and s are integers, the Eq. 5.1 can be modified as,

$$A(r,s) = e^{[\frac{-i\pi}{\lambda d}(r^2\Delta\xi^2+s^2\Delta\eta^2)]}$$
$$\times\Sigma_{k=0}^{N-1}\Sigma_{l=0}^{N-1} T(k,l) e^{\frac{-i\pi}{\lambda d}(k^2\Delta x^2+l^2\Delta y^2)}$$
$$\times e^{2i\pi(\frac{kr}{N}+\frac{ls}{N})} \tag{5.3}$$

In Eqn. 5.3, $A(r,s)$ is a matrix of NxN points that describes the amplitude and phase distribution of the real image and $\Delta\xi$ and $\Delta\eta$ are the pixel sizes in the reconstructed image of digital hologram. The Eqn. 5.3 is a representation of the Fresnel approximation in terms of a discrete Fourier transfor mation. This is an important fact, because the standard algorithms for a fast Fourier transformation can then be applied. In the experimental investigations a CCD array is placed at the position of the photosensitive surface (see Fig. 5.1). The CCD array consists of 1024×1024 light-sensitive pixels. The pixel area is $6.8\mu m \times 6.8\mu m$. The camera electronics produces a digital video output signal containing 256 gray levels per pixel. For computation the hologram is stored in a digital image processing system.

5.3 RECORDING ON CCD AND SAMPLING

In digital holography, recording the interefernce pattern between object and reference beam as an interference pattern on a CCD array and sampling it numerically is very important. This is because the hologram is recorded on CCD arrays and then stored in computer memory for numerical reconstruction. Since the hologram is a microscopically fine interference pattern generated by the coherent superposition of object and reference beam, the spatial frequency of it is defined by the angle between these two beams. Fig. 5.2 shows a geometry for recording digital Fresnel

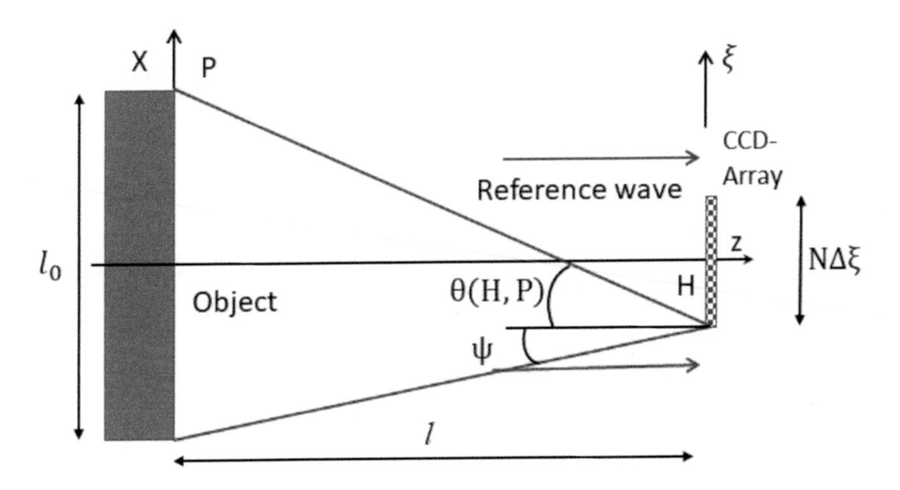

Figure 5.2 Experimental geometry for recording an in-line Fresnel digital hologram

hologram(in line) with CCD array which has $N \times M$ light sensitive pixels with a distance $\Delta\xi$ between pixel center along x direction and $\Delta\eta$ between pixel center along y direction. Assuming $N = M$ and $\Delta\xi = \Delta\eta$ and a plane reference wave incident on CCD normally as shown in Fig. 5.3, the distance d between two consecutive intereference fringes forming hologram at H is given by,

$$d = \frac{\lambda}{2sin(\frac{\theta}{2})} \tag{5.4}$$

The meaningful, sampling of intensity distribution for a hologram requires (Sampling theorem) that the period d is sampled with more than two pixels as, $d > 2\Delta\xi$ and this also can be expressed using spatial frequency ν of holographic finges as,

$$\nu < \frac{1}{2\Delta\xi} \tag{5.5}$$

In practice, CCD array will be specified by parameters $N, M, \Delta\xi, \Delta\eta$ which sets the limit for angle between reference and object beam θ and it will be small. In that case, $Sin\frac{\theta}{2} = tan\frac{\theta}{2} = \frac{\theta}{2}$ and the upper limit to angle $\theta < \frac{\lambda}{2\Delta\xi}$. The frequency ν becomes,

$$\nu < \frac{2}{\lambda} sin(\frac{\theta_{max}}{2}) \tag{5.6}$$

where, $\theta_{max} = max[\theta(H,P) : H,P]$. This clearly establishes that one can record holograms on a CCD array as long as the angle between object and reference beams is small and satisfies sampling theorem[6]. For the geometry shown in Fig. 5.2, for the object along lateral (x) length l_0, and with a plane reference beam propagating along optical axis onto the CCD, the maximum permissable object width l_0 can be calculated for each longitudinal distance of CCD from object surface l. As per Fig. 5.2 we have,

$$tan\theta = \frac{\frac{l_0}{2} + \frac{N\Delta\xi}{2}}{l} \tag{5.7}$$

substituting for the maximum angle θ we get,

$$\frac{\frac{l_0}{2} + \frac{N\Delta\xi}{2}}{l} < \frac{\lambda}{2\Delta\xi} \tag{5.8}$$

and the limit to lateral extension l_0 becomes,

$$l_0(l) < \frac{\lambda l}{\Delta\xi} - N\Delta\xi \tag{5.9}$$

from this we get value for minimum distance l required for recording digital hologram from the object to CCD plane as,

$$l(l_0) > \frac{(l_0 + N\Delta\xi)\Delta\xi}{\lambda} \tag{5.10}$$

For larger objects, each point of the object should have atleast one pixel satisfying sampling theorem which gives for angle ψ (Fig. 5.3),

$$tan\psi = \frac{\frac{l_0}{2} - \frac{N\Delta\xi}{2}}{l} \tag{5.11}$$

which modifies Eqn.5.10 for minimum object to recording plane distance as,

$$l(l_0) > \frac{(l_0 - N\Delta\xi)\Delta\xi}{\lambda} \tag{5.12}$$

Though in this case it is possible to obtain reasonably good hologram, for larger objects, the marginal points are stored only in few hologram points which results in weak contrast. In order to get appreciable contrast of digital holographic images the distance between object and CCD plane must satisfy,

$$l > l_0 \frac{\Delta\xi}{\lambda} \tag{5.13}$$

Thus it is important to note that while recording digital hologram the pixel size $\Delta\xi \times \Delta\eta$ plays a very crucial role in determining allowable angle between object and reference beam. If the pixel sizes of CCD or CMOS arrays available in marker are not comparable to resolution of a holographic plate, one can use at the recording plane an opaque screen consists of array of tarnsparent apertures(128) each of $2\mu m \times 2\mu m$ size with a pitch of $8\mu m$ in both directions[36]. The light transmitted through this apertures is magnified by a convex lens and then imaged on to CCD. Fig. 5.4 shows the geometry where, by shifting the mask using a piezo-translation stage in steps of $2\mu m$ along ξ, η directions a image composed of 16 sub-images can give a digital hologram of size $N\Delta\xi \times N\Delta\eta$ with values of $\Delta\xi = \Delta\eta = 2\mu m$ and $N = M = 4 \times 128 = 512$ respectively.

5.3.1 REDUCTION OF IMAGING ANGLE

In practice, we require for an object size of lateral dimension equal to 50 cm the distance between it and recording CCD array should be atleast 5.4 meter and this is large for recording a reasonably good digital hologram. To slove this one can use a concave lens(Fig. 5.3) where the object wave comes from a smaller virtual image of larger object. This geometry and use of virtual image of concave lens also satisfies

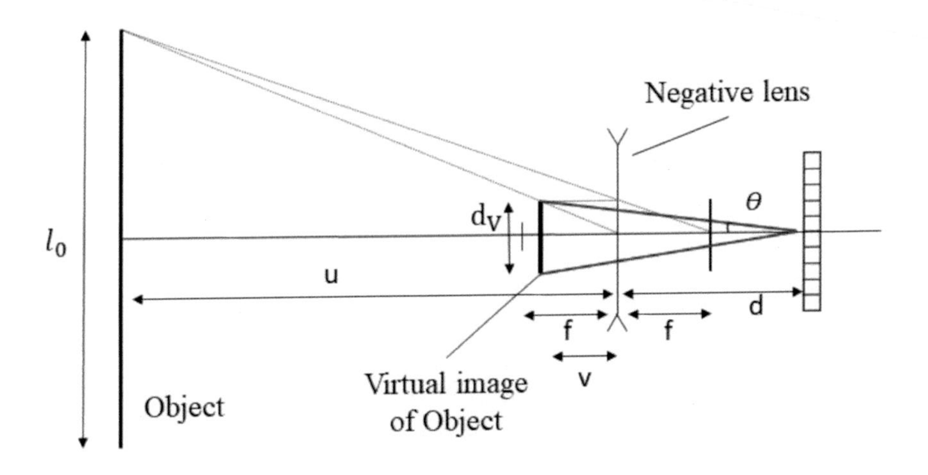

Figure 5.3 Geometry for reducing imaging angle in digital holography using a concave lens

($\theta < \frac{\lambda}{2\Delta\xi}$) the requirement of smaller angle between reference and object beam for recording digital hologram. Using the recording geometry shown in Fig. 5.3 we can directly calculate the distance d (between concave lens and CCD array) and the object distance u from lens as,

$$\frac{1}{f} = \frac{1}{u} - \frac{1}{v} \tag{5.14}$$

where, v is image distance from lens and the magnification is,

$$M_T = \frac{lv}{l_o} = -\frac{f}{u-f} \tag{5.15}$$

with M_T the translational magnification and l_v the lateral extension of virtual image. Simplifying further with values of $tan\theta = \frac{l_v}{[2(d+v)]}, l_v = \frac{l_o f}{(u-f)}$ and $v = \frac{uf}{(f-u)}$ using geometry shown in Fig. 5.4 we get,

$$d = \frac{-l_o f}{(u-f)2tan\theta} + \frac{uf}{(u-f)} \tag{5.16}$$

While reconstructing the hologram care must be taken to include the distance between CCD array and virtual image of object which is equal to $a = (v+d)$ instead of $(u+d)$. Also, the difference between lateral or longitudinal magnification or transverse magnification is also important when the object size is reduced due to insertion of lens. Apart from using concave lenses, convex lenses also can be used to reduce the size of object. Fig. 5.4 shows a typical digital holographic geometry using a

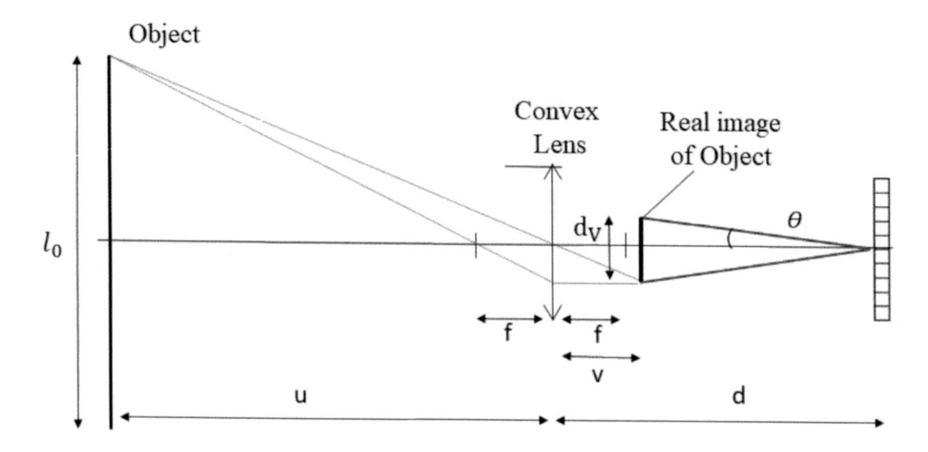

Figure 5.4 Geometry for reducing pixel size using a mask and convex lens in digital holography

convex lens instead of a concave lens. Then the distance d will become,

$$d = \frac{l_0 f}{(u-f)2tan\theta} + \frac{uf}{(u-f)} \tag{5.17}$$

where, focal length f is positive and the image is rael but upside down. In general, concave lenses are better as the total length $(u+d)$ is shorter and for digital holographic microscopy one can use convex lenses to magnify very small objects. Finally, instead of lenses apertures of appropriate dimensions also an be used to reduce the distance between object and CCD array but this limits the area of object size contributing to CCD pixel array which ultimately will reduce the contrast of hologram. The vignetting effect of aperture also contributes to reduction of whole field from object for recording on CCD pixels.

5.3.2 CONDITIONS FOR REFERENCE BEAMS

Unlike conventional holography, digital holography requires the reconstructing beam is accompanied by multiplication of digitally stored intensity distribution of object and reference beam with a digital model of reference beam and numerical determination of diffracted field in a partcular image plane. For achieving this different types of reference bemas can be used. i) If the reference beam is collimated one and collinear along with scaterred object beam then it will be easy for numerical reconstruction when it falls on a CCD array. Considering that the object is at x-y plane and the CCD array is kept at a distance z in ξ,η plane, then following[36] the numerical description of plane reference beam is,

$$R(\xi,\eta) = E_r + 0.0i \tag{5.18}$$

with E_r representing real distribution. For numerical evaluation $R(\xi,\eta) = 1.0$ becomes reference wave. Since the multiplication of the hologram with this reference wave does not change hologram, the plane normally impinging reference wave is the one used in digital holography. ii) In another method, Fig. 5.5 shows an oblique reference wave tilted by an angle with respect to (ξ,η) plane given by $\mathbf{k} = (\frac{2\pi}{\lambda})(0, sin\theta, cos\theta)$. At a point $R = (\xi,\eta,z)$ in hologram plane we have,

$$\begin{aligned} R(\xi,\eta) &= E_r e^{i(\mathbf{k}.\mathbf{r}+\psi)} \\ &= E_r e^{i(\frac{2\pi}{\lambda} zcos\theta+\psi)} e^{\frac{2\pi i}{\lambda}\eta sin\theta} \end{aligned} \tag{5.19}$$

in above equation 5.19, the first two exponential terms represent constant phase factor over hologram and the second exponential factor($R(\xi,\eta) = E_r e^{\frac{2\pi i}{\lambda}}\eta sin\theta)$ only represents the necessary plane reference beam impinging at an oblique angle θ. iii) One can also use a spherical wave diverging from a source point (x_s, y_s, z_s)(Fig. 5.6) and is given by,

$$R(\xi,\eta) = E_r e^{i(k.r+\psi)} \tag{5.20}$$

where $r = \sqrt{(\xi - x_s)^2 + (\eta - y_s)^2 + (d - z_s)^2}$ and $k = \frac{2\pi}{\lambda}$. In this configuration we can reconstruct the digital hologram using, with original diverging reference beam as well as using its convergent conjugate reference beam which is given by,

$$R(\xi,\eta) = E_r e^{i(k.r+\psi)} \tag{5.21}$$

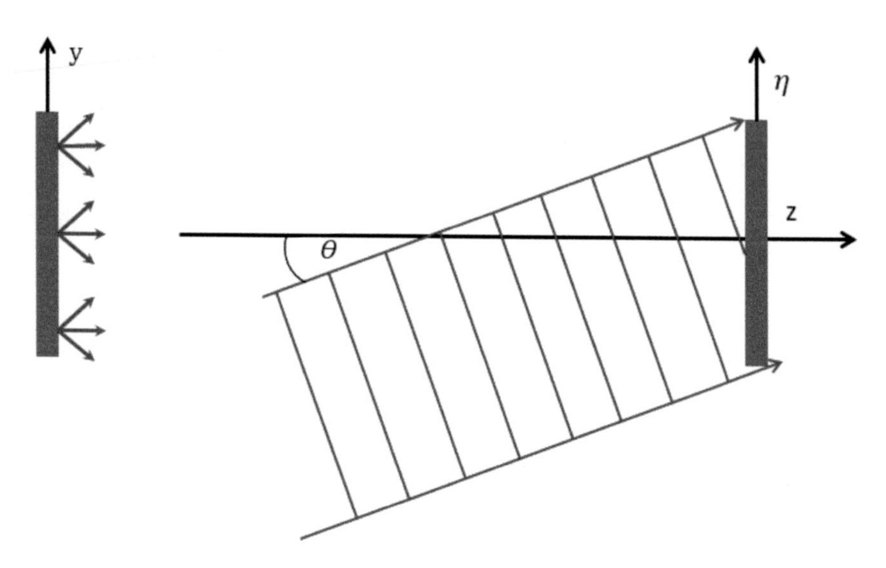

Figure 5.5 Oblique incidence plane reference beam illumination geometry for digital hologram

It may be noted here that if we multiply the digital hologram with original divergent reference beam, then we get a wave field diverging from virtual image in object plane and if we use conjugate reference beam then we get a wave field converging towards a real image in image plane. We can get lensless Fourier transform holography for a special case where $d = z_s$. Another important condition for resolving the micro interference using CCD array is that, the object and reference beams must be collinear. Fig. 5.7 shows a typical digital holographic geometry where, the object and reference beams are collinear and it uses a cube beamsplitter(CBS), which changes the optical path length between CCD aray and the object. Consider the refractive index of cube beam splitter is n_2, side length of CBS is l and the distance between CCD array and object is L as shown in Fig 5.7. The optical ray from object point O passes through the CBS and reaches the CCD array point D. This ray is bent due to at the beam splitter due to Snell's law ($n_1 sin\theta_1 = n_2 sin\theta_2$). Fig 5.7 explains clearly the ray diagram from object point to CCD array via the CBS. The angle of incidence is θ_1 and angle of refraction is θ_2 with $n_1 = 1$ the refractive index surrounding the CBS. The bent ray from object reaches CCD array and appears to be coming from point O' instead of O resulting in additional path length δ and is given by as per Fig. 5.7 as,

$$y = Ltan\theta_2 \tag{5.22}$$

$$y = (L - \delta)tan\theta_1 \tag{5.23}$$

and we get for δ,

$$\delta = \frac{(Ltan\theta_1 - Ltan\theta_2)}{tan\theta_1} = L(1 - \frac{1}{n_2}) \tag{5.24}$$

This δ value holds for all points of the object.

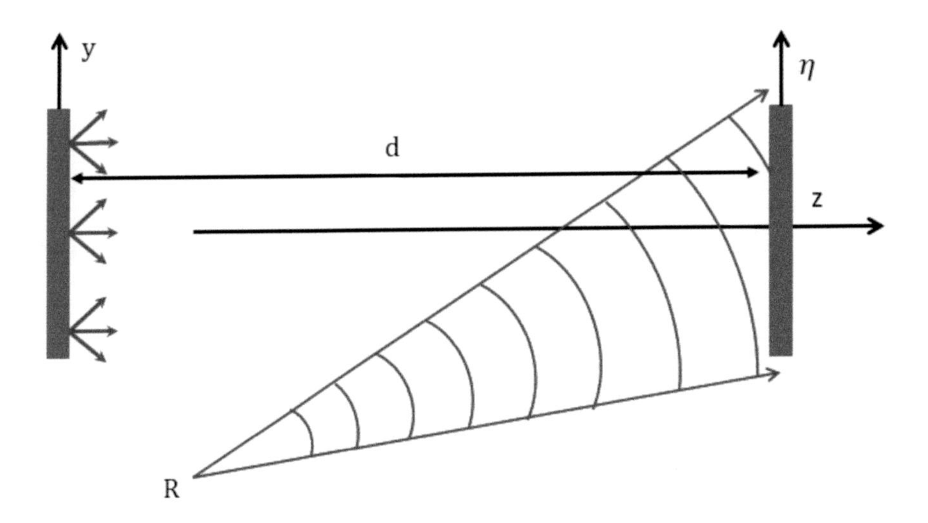

Figure 5.6 Spherical reference beam illumination geometry for digital hologram

5.4 NUMERICAL RECONSTRUCTION TECHNIQUES

5.4.1 INTRODUCTION

In this section we will be considering numerical reconstruction of object wave field from digitally recorded hologram on a CCD array. The theoretical explanation for numerical reconstruction is based on scalar diffraction theory[6]. Since the distance between object and CCD array is within the limit of Fresnel approximation, we will consider only the Fresnel approximation of diffraction integral. In this section the continuous formulas are transferred to finite discrete algorithms for implementation in digital image processing systems. Analysis about role of real and virtual images will be explored with respect to different types of reference waves. The main difficulty in numerical reconstruction is the suppression of D.C terms in the reconstructed image and the technique to suppress the same is presented in this section. We will describe two main numerical reconstruction techniques i) Finite discrete Frenel Transform method and ii) Convolution method.

5.4.2 RECONSTRUCTION USING FINITE DISCRETE FRESNEL TRANSFORM

Consider a diffusely reflecting object illuminated by a laser beam $E_{in}(x,y,z)|e^{j\psi(x,y,z)}$, and the corresponding complex reflected laser beam from the diffusely reflecting object will be,

$$b(x_0,y_0,z_0) = |b(x_0,y_0,z_0)|e^{j\phi(_0,y_0,z_0)} \tag{5.25}$$

Fig. 5.8 shows a typical numerical description geometry for digital holography where the object plane is represented by $b(x_0,y_0,z_0)$, the hologram plane by $h(\xi,\eta)$ and

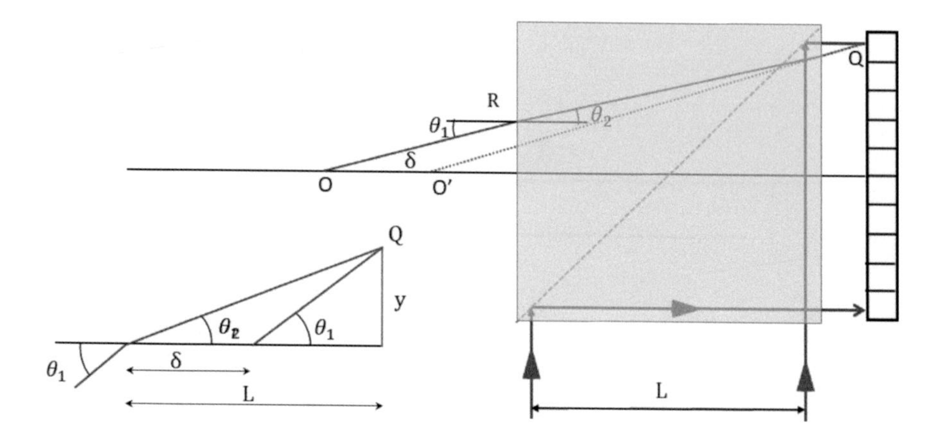

Figure 5.7 The digital holographic geometry using a cube beamsplitter.

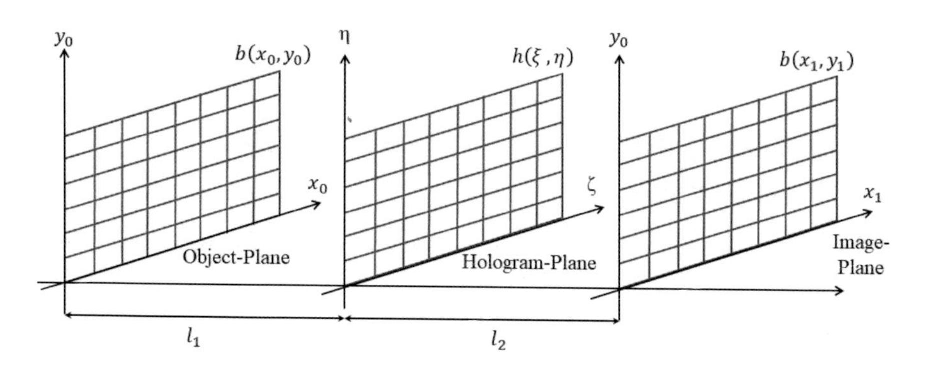

Figure 5.8 Geometry for digital holographic reconstruction

numerical reconstructing plane by $b(x_1, y_1)$ respectively. The distances between object plane to hologram plane is l_1 and hologram plane to image plane is l_2 respectively as shown in Fig. 5.8. Unlike conventional holography, the reconstruction plane in digital holography is image plane where,the real image will be numerically reconstructed. The object field distribution given by $E(\xi, \eta)$ and its superposition with reference beam $R(\xi, \eta)$ will interfere and give the intensity distribution at hologram plane $h(\xi, \eta)$. This intensity distribution consisting of superposition of object and reference beams is recorded by the CCD array. This intensity distribution is stored in a quantized and digitized form in the computer. The numerical reconstruction produces the complex distribution function $b(x_1, y_1)$ consisting the image of object. Using Fresnel's approximation the complex distribution at the hologram plane is given by,

$$E(\mu,\nu) = \frac{e^{jkl_1}}{j\lambda l_1} e^{j\pi l_1 \lambda(\mu^2+\nu^2)}$$
$$\int\int b(x_o,y_o) e^{\frac{j\pi}{l_1\lambda}(x_o^2+y_o^2)} e^{2j\pi(x_o\mu+y_o\nu)} dx_o dy_o \tag{5.26}$$

where, $\mu = \frac{\xi}{l\lambda}, \nu = \frac{\eta}{l\lambda}$ showing the relationship between co-ordinates (ξ,η) in hologram plane and the spatial frequencies $(\mu and \nu)$. We can neglect the factor $\frac{e^{jkl_1}}{j\lambda l_1}$ in the Eqn.5.26, as it does not depend on neither spatial frequency co-ordinates nor the object. In the hologram plane (ξ,η) the relation between pixel numbers N, M and pixel distances $\Delta\xi, \Delta\eta$ are given by,

$$\begin{aligned} \xi &= n\Delta\xi & n &= 1,2,3.....N \\ \eta &= m\Delta\eta & m &= 1,2,3.....M \end{aligned} \tag{5.27}$$

and in object plane the relationships are given by,

$$\Delta x_o = \frac{1}{N\Delta\mu} = \frac{l\lambda}{N\Delta\xi}, \Delta y_o = \frac{1}{M\Delta\nu} = \frac{l\lambda}{M\Delta\eta} \tag{5.28}$$

The digital hologram representing superposed reference wave $R(n\delta\xi, m\Delta\eta)$ and the object beam intensity $h(n\delta\xi, m\Delta\eta)$ at the plane $(\xi,\eta)$is then given by,

$$h(n\delta\xi, m\Delta\eta) = (E(\mu,\nu)+R)(E(\mu,\nu)+R)^* \tag{5.29}$$

To reconstruct the real and virtual images from the digitally stored hologram b_{re}, it is numerically multiplied with the reference beam. For reconstructing the real image the digital hologram b_{re} has to be multiplied with complex conjugate of reference beam $R^*(n\delta\xi, m\Delta\eta)$. Now, the real image will be obtained at the plane (x_i, y_i) due to the inverse Fresnel Transform. The real image can be viewed by putting $l_1 = l_2 = l$ and we obtain,

$$b_{Re}(x_1,y_1) = e^{\frac{i\pi}{\lambda l}(x_1^2+y_1^2)} \int\int h(\xi,\eta) R^*(\xi,\eta) e^{\frac{i\pi}{\lambda l}(\xi^2+\eta^2)} e^{\frac{-2j\pi}{l\lambda}(x_1\xi+y_1\eta)} d\xi d\eta \tag{5.30}$$

simplifying Eqn.5.30, further by substituting $\alpha = \frac{x_1}{l\lambda}, \beta = \frac{y_1}{\lambda l}$ we get,

$$b_{Re}(\alpha,\beta) = e^{i\pi l\lambda(\alpha^2+\beta^2)} \int\int h(\xi,\eta) R^*(\xi,\eta) e^{\frac{i\pi}{\lambda l}(\xi^2+\eta^2)} e^{-2j\pi(\alpha\xi+\beta\eta)} d\xi d\eta \tag{5.31}$$

Eqn.5.31 clearly shows that the field of real image is the Fourier Transform of digital hologram $h(\xi,\eta)$ multiplied by conjugate of reference beam $R^*(\xi,\eta)$ and the function $e^{\frac{i\pi}{\lambda l}(\xi^2+\eta^2)}$. The discrete version of Eqn.5.30 is given by,

$$b_{Re}(n\Delta\alpha, m\Delta\beta) = e^{i\pi l\lambda(n^2\Delta\alpha^2+m^2\Delta\beta^2)}$$
$$\Sigma_{p=0}^{N-1}\Sigma_{q=0}^{M-1} h(p\Delta\xi, q\Delta\eta) R^*(p\Delta\xi, q\Delta\eta) e^{\frac{i\pi}{\lambda l}(p^2\Delta\xi^2+q^2\Delta\eta^2)}$$
$$e^{-2\pi j(\frac{pn}{N}+\frac{qm}{M})} \tag{5.32}$$

In Eqn.5.32 $\Delta\alpha = \frac{1}{N\Delta\xi} = \frac{\Delta x_1}{l\lambda}$ and $\Delta\beta = \frac{1}{M\Delta\eta} = \frac{\Delta y_1}{l\lambda}$ respectively. The pixel spacing in real digital holographic image coincides with the corresponding pixel spacing in the object plane with $\Delta x_1 = \Delta x_o, \delta y_1 = \Delta y_o$. With these explanations finally we get the main reconstruction formula for digital holography equal to,

$$b_{Re}(n\Delta x_1, m\Delta y_1) = e^{i\pi l\lambda(\frac{n^2}{N^2\Delta\xi^2}+\frac{m^2}{M^2\Delta\eta^2})}$$
$$\Sigma_{p=0}^{N-1}\Sigma_{q=0}^{M-1}H(p\Delta\xi, q\Delta\eta)R^*(p\Delta\xi, q\Delta\eta)e^{\frac{i\pi}{\lambda l}(p^2\Delta\xi^2+q^2\Delta\eta^2)}$$
$$e^{-2\pi j(\frac{pn}{N}+\frac{qm}{M})} \tag{5.33}$$

If the reference wave component $R(\xi, \eta)$ is incident normally (Equal to 1), then multiplication of $R^*(p\Delta\xi, q\Delta\eta)$ in Eqn.5.33 can be neglected. Eqn.5.33 represents final practical method to numerically reconstruct a digital hologram using Fresnel Transform method and $b_{Re}(n\Delta x_1, m\Delta y_1)$ represents a complex object wave field with intensity is given by,

$$I(n\Delta x_1, m\Delta y_1) = |b_{Re}(n\Delta x_1, m\Delta y_1)|^2 \tag{5.34}$$

and the phase of object field is equal to,

$$\phi(n\Delta x_1, m\Delta y_1) = arctan\frac{Im[b_{Re}(n\Delta x_1, m\Delta y_1)]}{Re[b_{Re}(n\Delta x_1, m\Delta y_1)]} \tag{5.35}$$

Thus, unlike conventional optical holography or photorefractive dynamic holography the phase of object field can be reconstructed which is very important for holographic interferometeric applications. Now, the pixel sizes are equal to $\Delta x_i = \frac{l\lambda}{N\Delta\xi}$ and $\Delta y_1 = \frac{l\lambda}{M\Delta\eta}$ respectively. For this relation it is possible to have the exponential term of Eqn.5.33 in the form as $e^{2\pi j(\frac{pn}{N}+\frac{qm}{M})}$ which enables us to use FFT algorithm. The reconstructed digital holographic images are defined by the numbers N and M and dsitances of pixels $\Delta\xi, \Delta\eta$ respectively. The reconstructed digital holographic image using Fresnel reconstruction method has field of size,

$$N\Delta x_o \times M\Delta y_o = \frac{l\lambda}{\Delta\xi} \times \frac{l\lambda}{\Delta\eta} \tag{5.36}$$

which is normally displayed in full frame of monitor of image processing system. It can be noted that pixel distances $\Delta x_1 = \Delta x_o$ and $\Delta y_1 = \Delta y_o$ depend upon wave length of light beam used λ and reconstruction distance $l_1 = l_2 = l$.

5.4.2.1 Reconstruction of Real and Virtual image

We know that the reconstruction process in conventional holography, uses a reference wave and it can yield both real and virtual images of hologram simultaneously. The real image is reconstructed due to a converging beam from object and real image is due to diverging beam emitted from the object during recording of hologram. To reconstruct the virtual image, a convex lens is required which is often observer's eye or objective lens of recording camera. In case of digital holography, when the recorded digital hologram is multiplied numerically with a reference beam, the converging beam emerging from object intensity produces sharp real image along with unsharp virtual image produced due to diverging beam emerging from object intensity respectively at the image plane $l_1 = l_2 = l$. To reconstruct a sharp virtual image numerically consider a convex lens of focal length is introduced behind the hologram plane. Then using len's equation $\frac{1}{f} = \frac{1}{l_1} + \frac{1}{l_2}$ and assuming unit magnification, we get $f = \frac{l}{2}$ and due to that the phase transformation by the lens becomes,

$$T_v(\xi,\eta) = e^{[\frac{-2j\pi}{\lambda l}(\xi^2+\eta^2)]} \tag{5.37}$$

and the field in the image plane (x_i, y_i) will become,

$$H(\xi,\eta)R^*(\xi,\eta)e^{-\frac{2\pi j}{\lambda l}(\xi^2+\eta^2)} \tag{5.38}$$

Now, with this modification due to chirp function $e^{-\frac{2\pi j}{\lambda l}(\xi^2+\eta^2)}$ Eqn.5.33 will become,

$$H(p\Delta\xi,q\Delta\eta)R^*(p\Delta\xi,q\Delta\eta)e^{-\frac{2\pi j}{\lambda l}(p^2\Delta\xi^2+q^2\Delta\eta^2)}e^{\frac{\pi j}{\lambda l}(p^2\Delta\xi^2+q^2\Delta\eta^2)} \tag{5.39}$$

Simplifying above Equation 5.39 we get,

$$= H(p\Delta\xi,q\Delta\eta)R(p\Delta\xi,q\Delta\eta)e^{\frac{-\pi j}{\lambda l}(p^2\Delta\xi^2+q^2\Delta\eta^2)} \tag{5.40}$$

Equation 5.40 shows that virtual image can be obtained by putting $l = -l$ and the intensity of virtual image can be obtained at the object plane as $z = -l$. Normally there will not be any significant difference between real and virtual images except for 180° rotation which looks like twin images.

5.4.2.2 The D.C Term of Fresnel Transform

One of the important terms which we see while reconstructing numerically is the bright central square for any object known as D.C term. This will be more brighter than the reconstructed real or virtual images of recorded object. This central bright spot is nothing but zeroth order diffraction pattern and it is undiffracted part of reconstructing reference wave. In computational point of view this is nothing but D.C term of Fresnel Transform. IThe factors before integarls in Eqn.5.30 or before the sums in Eqn.5.33 only will affect the phase and is independent of specific hologram. Since Fresnel transform is Fourier transform of a product the abive said factors is hologram times reference wave i.e $H.R^*$ and the chirp function and according to convolution theorem this process gives same as convolution of Fourier Transform of individual factors. Then $F.T[H(p\Delta\xi,q\Delta\eta)].R^*[(p\Delta\xi,q\Delta\eta)]$ gives trimodal with a high amplitude peak at the spatial frequency $(0,0)$. This is D.C term represented by $H(0,0)$,and it is equal to,

$$H(0,0) = \Sigma_{p=0}^{N-1}\Sigma_{q=0}^{M-1}H(p\Delta\xi,q\Delta\eta)R^*(p\Delta\xi,q\Delta\eta) \tag{5.41}$$

The D.C term $H(0,0)$ is also can be considered as Dirac-Delta function. Since the D.C term of Fresnel transform is same as D.C term of Fourier Transform of digital hologram multiplied by convolution of reference wave with chirp function, we can represent D.C term of Fresnel Transform as Fourier Transform of Chirp function and is equal to,

$$e^{\frac{\pi j}{\lambda l}(p^2\Delta\xi^2+q^2\Delta\eta^2)} = e^{\frac{\pi j}{\lambda l}p^2\Delta\xi^2} e^{\frac{\pi j}{\lambda l}q^2\Delta\eta^2} \tag{5.42}$$

and it is restricted to finite extent of hologram. In two dimensions the area of D.C term is equalent to,

$$= \frac{N^2\Delta\xi^2}{l\lambda} \times \frac{M^2\Delta\eta^2}{l\lambda} \tag{5.43}$$

where, $\frac{N^2\Delta\xi^2}{l\lambda}$ represents width of D.C term along x_i direction and $\frac{M^2\Delta\eta^2}{l\lambda}$ represents width of D.C term along y_i direction. If the pixel dimensions increases along with pixel number of CCD target then the width of D.C term also increases but decreases for increasing distance of l. Also, a shift in finite chirp function changes the location of its Fourier Transform and the D.C term. Consider that we are reconstructing a digital hologram using Eqn. 5.33, then it is defined as $[0, N\Delta\xi] \times [0, M\Delta\eta]$ and the chirp function given by $e^{\frac{\pi j}{\lambda l}p^2\Delta\xi^2} e^{\frac{\pi j}{\lambda l}q^2\Delta\eta^2}$ carries the local frequencies from 0 to $\frac{N\Delta\xi}{l\lambda\pi}$ along ξ direction and from 0 to $\frac{M\Delta\eta}{l\lambda\pi}$ along η direction respectively. Thus the square D.C term will appear at the center of digital holographic intensity pattern. Now, if we shift the chirp function by $p_0\Delta\xi$ along ξ direction and $q_0\Delta\eta$ along η direction then we get,

$$= e^{[\frac{\pi j}{\lambda l}(p-p_0)^2\Delta\xi^2]} e^{[\frac{\pi j}{\lambda l}(q-q_0)^2\Delta\eta^2]} \tag{5.44}$$

The terms shown in above Eqn. 5.44 carries local frequencies from $\frac{-p_0\Delta\xi}{l\lambda\pi}$ to $\frac{(N\Delta\xi-p_0\Delta\xi)}{(l\lambda\pi)}$ and from $\frac{-q_0\Delta\eta}{l\lambda\pi}$ to $\frac{(M\Delta\eta-q_0\Delta\eta)}{(l\lambda\pi)}$ along two directions ξ and η respectively.

5.4.2.3 Suppression of the D.C Term

The D.C term in digital holography is similar to the zero order image in conventional holography. The twin image(Real and virtual) is similar to object and conjugate images in conventional holography. For in-line Gabor holography the zeroth order and conjugate image posed similar problem and Leith and Upatnieks off-axis holography solved this twin image problem to isolate holographic image of the original object. Thus it is necessary to suppress the D.C term of digital holography and solve twin image problem. In digital holography using effective numerical methods one can effectively suppress D.C term disturbing real image of digital holography. Though there are several methods reported in the literature for suppression of D.C term, the simplest numerical method to suppress is developed by Kries and Juptner[37]. Following Kries and Juptner[37], consider in a digital holography reconstruction process, using the recorded digital hologram ($H(p\Delta\xi, q\Delta\eta)$ multiplied with reference beam $R^*(p\Delta\xi, q\Delta\eta)$, we find the average intensity as,

$$H_{av} = \frac{1}{NM}\Sigma_{p=0}^{N-1}\Sigma_{q=0}^{M-1} H(p\Delta\xi, q\Delta\eta)R^*(p\Delta\xi, q\Delta\eta) \tag{5.45}$$

subtracting average intensity H_{av} from each stored hologram intensity value we get the modified digital hologram as given by,

$$H_i(p\Delta\xi, q\Delta\eta) = H(p\Delta\xi, q\Delta\eta)R^*(p\Delta\xi, q\Delta\eta) - H_{av}$$
$$with p = 0.......N-1; q = 0..........M-1. \tag{5.46}$$

since the D.C term in Fourier spectrum of $H(0,0)$ calculated as per Eqn.5.41 is zero the convolution of a zero D.C term with transform of chirp function becomes zero. Thus we can efficiently suppress the D.C term in the modified hologram H_i. This process of suppressing a D.C term in digital holography is equal to application of a high pass filter with a cut off frequency equal to smallest non-zero frequency with suppression of spatial frequency $(0,0)$. Thus one can employ other high pass filters for suppressing small spatial frequencies to achieve comparable results. Subtracting averages over each 3×3 pixels in neighborhood from the original digital hologram we get,

$$H_i(p,q) = H(p,q) - \frac{1}{9}[H(p-1,q-1) + H(p-1,q) + H(p-1,q+1) +$$
$$H(p,q-1) + H(p,q) + H(p,q+1) + H(p+1,q-1) +$$
$$H(p+1,q) + H(p+1,q+1)], p = 2......N-1; q = 2........M-1. \tag{5.47}$$

In Eqn.5.47 the terms $\Delta\xi, \Delta\eta$ are omitted in the pixel arguments. For the intensity of hologram given by $H_o(x,y) = I_o + Ir = \sqrt{I_o I_r} cos(kr + \phi)$ the sum of the squared sines becomes zero for zero frequency and slowly varying. This suppresses zero and low frequency terms leaving only $+1$ and -1 diffraction orders representing real and virtual images which are separated well from D.C term.

5.4.2.4 Suppression of twin images in digital holography

Another important suppression required to obtain clear digital holographic image is to eliminate twin images. To do that the object to be digitaly holographed is placed outside the optical axis such that the twin images do not overlap while numerically reconstructing. This will result in two well separated distinct partial spectra of the amplitude spectrum representing real and virtual images respectively. By setting the virtual image spectrum zero we can clearly reconstruct real image of digital hologram. There are several methods described in literature for suppressing twin images and one must be careful to choose appropriate one and the elimination of twin images is must to get high efficient digital hologram.

5.4.3 NUMERICAL RECONSTRUCTION OF DIGITAL HOLOGRAM BY CONVOLUTION METHOD

5.4.3.1 Diffraction integral as a convolution

The digital hologram reconstruction process can also be done using convolution theorem approach. This is because the diffraction formula is the superposition integral of a linear shift invariant system with which we can apply convolution technique for recosntruction of digital hologram. Using convolution integral the diffracted field at the hologram plane is given by,

$$H_i(x_i,y_i) = \int\int H(\xi,\eta).R^*(\xi,\eta)G(x_i-\xi,y_i-\eta)d\xi d\eta \tag{5.48}$$

The function $G(x_i-\xi,y_i-\eta)$ is impulse response function due to free space propagation and is given by,

$$G(x_i-\xi,y_i-\eta) = \frac{j}{\lambda}\frac{e^{j\sqrt{(x_i-\xi)^2+(y_i-\eta)^2+l^2}}}{\sqrt{(x_i-\xi)^2+(y_i-\eta)^2+l^2}} \tag{5.49}$$

The eqn.5.48 can be written as,

$$H_i(x_i,y_i) = [H(\xi,\eta).R^*(\xi,\eta)] * G(\xi,\eta) \tag{5.50}$$

where * implies convolution operation. This process of using convolution theorem reduces significantly the computaion required for the reconstruction by replacing the convolution in spatial domain by a multiplication of the complex spectra in spatial frequency domain followed by inverse Fourier Transform of this product into spatial domain. Using Fast Fourier Transform algorithm we can calculate forward Fourier Transform F as well as its inverse Fourie Transform F^{-1}. Now, the Eqn.5.48 becomes,

$$H_i = F^{-1}[F(H.R^*).F(G)] \tag{5.51}$$

Replacing the continuous co-ordinates (ξ,η) with discrete values $p\Delta\xi$ and $q\Delta\eta$ respectively the impulse response function $G(p,q)$ can be written as,

$$G(p,q) = \frac{j}{\lambda}\frac{e^{\frac{2\pi j}{\lambda}\sqrt{(p-1)^2\Delta\xi^2+(q-1)^2\Delta\eta^2+l^2}}}{\sqrt{(p-1)^2\Delta\xi^2+(q-1)^2\Delta\eta^2+l^2}} \tag{5.52}$$

we can do the programming by using $je^{j\theta} = (-sin\theta + jcos\theta)$ and the discrete impulse response $G(p,q)$; $p = [1,......,N], q = [1.........M]$ has to be transformed via FFT algorithm to obtain the discrete transfer function G and is given by,

$$\begin{aligned} G(\mu,\nu) &= e^{(\frac{-2\pi j}{\lambda})\sqrt{1-(\lambda\mu)^2-(\lambda\nu)^2}} : (\lambda\mu)^2+(\lambda\nu)^2 \leq 1 \\ G(\mu,\nu) &= 0 \qquad : otherwise \end{aligned} \tag{5.53}$$

Where the discrete values ofμ and ν are given by,

$$\mu = \frac{(m-1)}{(M\Delta\xi)} \quad and \quad \nu = \frac{(n-1)}{(N\Delta\eta)} \tag{5.54}$$

where $m = 1.....M$, and $n = 1...........N$ respectively. Now, finite discrete function becomes,

$$G(m,n) = e^{(\frac{2\pi jl}{\lambda})\sqrt{1-(\frac{\lambda(m-1)}{M\Delta\xi})^2-(\frac{\lambda(n-1)}{N\Delta\eta})^2}}$$

$$G(m,n) = e^{(\frac{2\pi jl}{M\Delta\xi})\sqrt{\frac{M^2\Delta\xi^2}{\lambda^2}-((m-1)^2-(n-1)^2}} \tag{5.55}$$

The last eqn.5.55 holds for frequency occuring at $M = N$ and $\Delta\xi = \Delta\eta$. In general we have four procedures to reconstruct digital hologram using convolution approach. We can either define exact impulse response $G(p,q)$ or its Fresnel approximation $G_F(p,q)$ and find $F.T[G]orF.T[G_F]$ and in a similar way we can use the transfer function $G(m,n)$ or its Fresnel approximation $G_f(m.n)$. Using any of these four expressions we can reconstruct the digital hologram using $H_i = F^{-1}(F[H].G)$.

5.4.3.2 Image Field size in convolution approach

In case of Fresnel Transform reconstruction approach if the hologram plane (ξ,η) is defined as spatial domain then it will become Fourier spectrum of a product in spatial domain. Now, the resolution becomes $\Delta x_i = \frac{d\lambda}{(N\Delta\xi)}$ in the spatial domian at the plane of real image and same holds good for virtual image. On the other hand, in case of the convolution approach the digitally recorded hologram on a CCD multiplied with reference beam is tranformed from spatial domain to spatial frequency domain by multiplying the spectrum with the transfer function. Then by taking inverse Fourier Transform this product is tranferred back into spatial domain. This process results in a reconstructed image with resolution similar to original hologram with M values with spacing $\Delta\xi$ and N values with spacing $\Delta\eta$ respectively. This will result in same size and resolution of holographic image unlike in Fresnel Transform method where the reconstructed digital holographic image depends upon either reconstruction distance or wavelength λ. This makes convolution based numerical reconstruction method suitable for Gabor type in-line holography. If the object field has a width equal to $N\Delta\xi$ of CCD array then convoloution reconstruction method will reconstruct the image to full field. In case of Fresnel Transformation approach at the image field the width will be,

$$N\Delta x_i = \frac{Nd\lambda}{N\Delta\xi} = \frac{d\lambda}{\Delta\xi} \tag{5.56}$$

and for in-line digital holography the width becomes,

$$N\Delta x_i = \frac{Nd\lambda}{N\Delta\xi} = \frac{N\Delta\xi^2}{d\lambda} \tag{5.57}$$

which is equal to width of D.C term in the reconstruction. But in convolution method the rigidity of fixed and limited size of reconstructed image in image plane poses problems for larger sized objects having lateral extent more than the CCD array.

When we reconstruct numerically the image in convolution method, using either impulse response function $G(x_i - \xi, y_i - \eta)$(Eqn.5.49) or the transfer function $G(\mu, \nu)$, the reconstructed size of image will have same size as well as location of D.C term. For reconstructing part of the images besides D.C term from this the origin of image plane has to be shifted as the system remains shift invariant. Now, the shifted version of impulse response becomes,

$$G_s(x_i - \xi, y_i - \eta) = \frac{i}{\lambda} \frac{e^{jk\sqrt{(x_i-\alpha-\xi)^2+(y_i-\beta-\eta)^2+l^2}}}{\sqrt{(x_i-\alpha-\xi)^2+(y_i-\beta-\eta)^2+l^2}} \tag{5.58}$$

The terms α, β in above equation represents shift along x_i, y_i directions respectively. Finally, the discrete finite version with shifts s_p, s_q becomes,

$$G(p+s_p, q+s_q) = \frac{i}{\lambda} \frac{e^{jk\sqrt{(p+s_p)^2(\Delta\xi)^2+(q+s_q)^2(\Delta\eta)^2+l^2}}}{\sqrt{(p+s_p)^2(\Delta\xi)^2+(q+s_q)^2(\Delta\eta)^2+l^2}} \tag{5.59}$$

where $k = \frac{2\pi}{\lambda}$. Reconstruction of impulse response function shown in above Eqn.5.59 results in shifting of field by a value equalent to $(s_p\Delta\xi, s_q\Delta\eta)$. There are several reconstruction methods based on possible field shifts (s_p, s_q) are given in literature[36]. Further we can control the size of digitally reconstructed image to fit the size of object. This is because in case of Fresnel Transform approach, the reconstructed image size depends upon the distance l, wavelength λ, pixel size $\Delta\xi \times \Delta\eta$ and the pixel number $M \times N$ and in case of convolution approach, the image field coincides with CCD dimensions given by $M\Delta\xi \times N\Delta\eta$. While digitally reconstructing a small die (object) it can occupy larger image field area when Fresnel Transform method is used and on the other hand if, convolution method of numerical reconstruction is used then the image field will cover only small area of object size due to CCD array limitation. This necessiates the need for controlling the digital image size of the reconstructed image field to exactly fit in to original object size. In the case of Fresnel reconstruction process, a change in either distance l or wavelength λ results in blurred unsharp images. Instead if we use the convolution approach, by rescaling the size of digitally stored hologram we can exactly fit the image field similar to object size. To carry out rescaling of hologram, the $M \times N$ pixel hologram is broadened to $2M \times 2N$ pixel hologram. Previously, the original $M \times N$ is surrounded by pixels of intensity 0 and since now the pixel numbers in each direction are doubled to $2N, 2M$ respectively, the pixel size $\Delta\xi \times \Delta\eta$ remains same. This gives a reconstructed image of size equal to $4MN\Delta\xi, \Delta\eta$. For keeping the same numbers M and N while doubling the image size in each direction, and create new hologram with $\frac{M}{2} \times \frac{N}{2}$ pixels with a pixe size of $\Delta\xi_1 = 2\Delta\xi$ and $\Delta\eta_1 = 2\Delta\eta$. Now this new hologram results in central part of $M \times N$ pixel hologram with black pixels surrounding it. Further, the numerical recosntruction yileds filed size of $4MN\Delta\xi, \Delta\eta = M\Delta\xi_1 \times N\Delta\eta_1$. One can slao combine both approaches to scale the reconstructed image field by reducing original hologram $(M\Delta\xi \times N\Delta\eta)$ to a $(2\Delta\xi, 2\Delta\eta, 2N, 2M)$ hologram which will give a reconstructed image field size equal to $16NM\Delta\xi, \Delta\eta = 2M2\Delta\xi \times 2N2\Delta\eta$. This will contain pixels of size 2048×2048 of $13.6\mu m^2$ each with only pixels of size

512×512 occupying central portion of recoded hologram. The image field sizes in digital holographic reconstruction can be rescaled using a different reference beam compared to the one used, while recording the digital hologram. This is because the change of reference wave geometry using non-collimated beams or change of reference wave wavelength while reconstruction it, can give lateral magnification. This lateral magnification due to change in reference wave results in same pixel size magnification similar to one which we have described earlier in the first chapter. Fig.

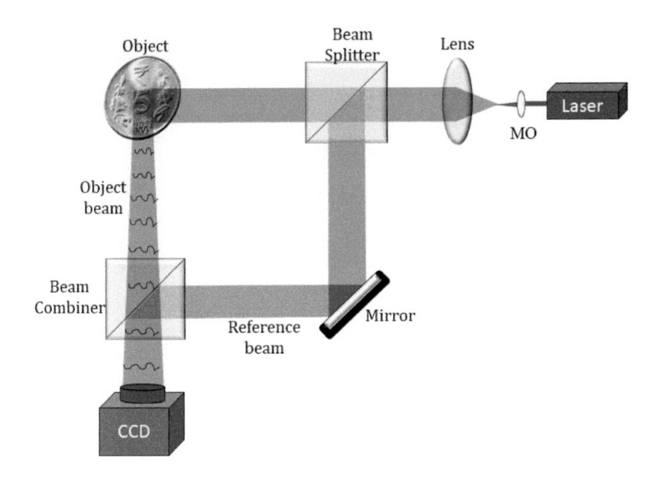

Figure 5.9 Digital holographic set-up in laboratory for recording digital hologram of a coin.

5.9 shows a typical experimental geometry for recording a coin in digital holography where the laser beam is split in to two parts by a beamsplitter. One of beam illuminates the object and the other beam is reference in an off-axis digital holographic set up as shown in Fig. 5.9. The hologram is recorded on the CCD camera(Fig. 5.9) and after numerical reconstruction we get the digital holographic image of the coin object as shown in Fig. 5.10. During the reconstruction process Fresnel Transform approach using matlab program was used. This simple matlab algorithm for digital holographic reconstruction can be done by any undergraduate student of physics or electronics engineering.

5.5 PHASE SHIFTING DIGITAL HOLOGRAPHY

5.5.1 INTRODUCTION

We discussed extensively the procedure for recording digital holography in previous sections including the reconstruction process. To record digital holography the of-axis recording methods are used and in that, a complex field in hologram plane was generated by multiplying the recorded digital hologram with the reference wave. For reconstruction, the field in image plane was calculated by simulating free space

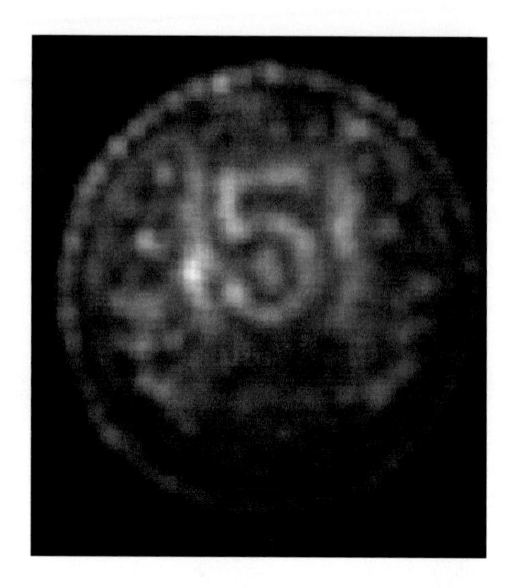

Figure 5.10 Numerically reconstructed digital hologram of 5 Rupees coin.

propagation using either Fresnel transform or convolution approach. The multiplication of real digital hologram of an object with complex reference wave gives three components in the reconstructed field namely, the objec field(real image), its conjugate field(virtual image) and the d.c term respectively. The recording methods used in digital holography has an off-axis hologram geometry, which prohibits effective use of the pixel number of a CCD because of the necessity for carrier fringes and also the size of the reconstructed image is limited by the presence of zero-order and conjugate images. One can use an in-line hologram free from these limitations but, it requires digital filtering to suppress the conjugate image. For surface mapping of 3-D objects, the technique of wavelength-scanning holography with a tunable dye laser can be used, but it requires an especially high computation load. To solve such difficulties Yamaguchi and Zhang[38] proposed a new digital holography technique by phase-shifting the reference wave with respect to object beam. For phase shifting the reference beams, in general we have two major techniques of phase shifting i) Dynamic phase shifting and ii) Geometric phase shifting used in interferometry. The dynamic phase shifting method which was used in phase shifting interferometry and can be obtained by using a rotating half wave plate, shifted diffraction grating, tilted glass plate, piezo-electric mirror and elongation of fiber[36]. Another way of phase shifting is geometric phase shifting, first introduced by Chyba et.al[39] in a Michelson interferometer. The process of phase shifting, measures the complex amplitude of the object wave at the CCD plane located at finite distance in the in-line setup. Yamaguchi and Zhang[38,40] are the first to use a dynamic phase shifting of reference wave in 4 steps to record a digital hologram and later applied it to microscopy.

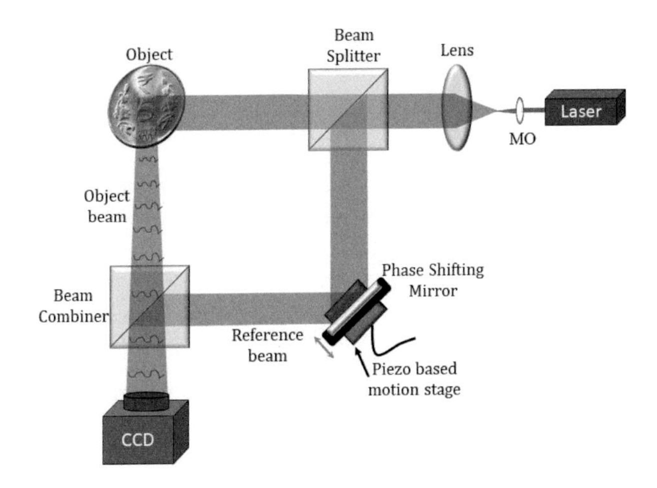

Figure 5.11 Experimental geometry for recording dynamic phase shifting digital hologram

5.5.2 DYNAMIC PHASE SHIFTING DIGITAL HOLOGRAPHY

A typical holographic geometry with a piezo electric mirror replacing conventional mirror is shown in Fig. 5.11. The light beam from laser source is divided in to two parts and one of them falls on the object to be recorded and the other one(reference) is reflected by a piezo-electric mirror connected to a computer. The phase shifting of reference beam is done by the piezo-controller connected to the mirror. Consider (x_o, y_o, z_o) is the object position, (ξ, η, z) represents the hologram plane and (x_i, y_i, z_i) represents the image plane respectively. The complex object field at the hologram plane is given by,

$$U_h(\xi,\eta) = Ue^{j\phi} \quad (5.60)$$

$$= \frac{U_0}{z_0} e^{[j\phi_0 + jkz_0 + jk[\frac{(\xi - x_0)+(\eta - y_o)}{2z_o}]} \quad (5.61)$$

If $U_R e^{j\phi_r}$ represents the reference wave then the total intensity at the hologram plane will be,

$$I_H = |U_h(\xi,\eta) + U_R e^{j\phi_R}|^2 \quad (5.62)$$

$$I_H = U_h^2 + U_R^2 + 2U_h U_o cos(\phi_o - \phi_R) \quad (5.63)$$

Now, the reference eam is phase shifted by $0, \frac{\pi}{2}, \pi, \frac{3\pi}{2}$ to get the four intensity patterns as,

$$\begin{aligned} I_H(\xi,\eta) &= I_h + I_R + 2\sqrt{I_h I_R} cos\phi_o \\ I_H(\xi,\eta) &= I_h + I_R + 2\sqrt{I_h I_R} sin\phi_o \\ I_H(\xi,\eta) &= I_h + I_R - 2\sqrt{I_h I_R} cos\phi_o \\ I_H(\xi,\eta) &= I_h + I_R - 2\sqrt{I_h I_R} sin\phi_o \end{aligned} \quad (5.64)$$

From this we get for object phase information ϕ_o equal to,

$$\phi_o(\xi,\eta) = tan^{-1}\frac{I_H(\xi,\eta,\frac{\pi}{2}) - I_H(\xi,\eta,\frac{3\pi}{2})}{I_H(\xi,\eta,0) - I_H(\xi,\eta,\pi)} \tag{5.65}$$

In arriving above relation we have assumed the initial reference phase as 0 and we can reconstruct numerically the object field at image plane (x_i, y_i) as shown in Fig. 5.12 using Fresnel transformation approach as,

$$U_i(x_i,y_i,z_i) = \int\int U_h(\xi,\eta)e^{\frac{jk(x_i-\xi)^2+(y_i-\eta)^2}{2z_i}} d\xi, d\eta \tag{5.66}$$

Substituting for $U_h(\xi,\eta)$ from Eqn.5.61 and rearranging the terms we get,

$$U_i(x_i,y_i,z_i) = \frac{U_0}{z_0}e^{[j\phi_0+jkz_0+\frac{jk}{2}\frac{(x_i^2+y_i^2)}{z_i}+\frac{(x_o^2+y_o^2)}{z_o}]}$$
$$\times\int\int e^{-jk[\xi(\frac{x_o}{z_o}+\frac{x_i}{z_i})+\eta(\frac{y_o}{z_o}+\frac{x_i}{z_i})-\frac{(\xi^2+\eta^2)}{2}(\frac{1}{z_o}+\frac{1}{z_i})]} d\xi d\eta \tag{5.67}$$

and the intensity of $|U_i(x_i,y_i,z_i)|^2 = I_i(x_i,y_i,z_i)$ becomes maximum at $x_i = x_o, y_i = y_o, z_i = z_o$ indicating the real image at the original object position. Yamaguchi and Zhang(Ref) in this experiment have used an Argon-Ion laser of wavelength equal to 514.5 nm with an 1 watt output power as source and the reference mirror used was moved in steps to introduce phase shifting a by computer controlled piezo-electric actuator. The camera used to record digital hologram was Sony-XC-77 CCD with 493×768 pixels of $13\mu m$ size and the out put was stored in a frame grabber with 512×512 pixels. Fig. 5.12 shows the image of a coin object recorded using dynamic phase shifting digital holography using the experimental geometry shown in Fig. 5.11. Now, this phase shifting digital hologram can be implemented in any graduate optics laboratory using a CMOS/CCD camera and personal computer with frame grabber faciliy.

5.5.3 QUADRATURE DYNAMIC PHASE SHIFTING DIGITAL HOLOGRAPHY IN TWO STEPS

Gabor and Goss proposed a quadrature phase-shifting holography (QPSH), also called two-step phase-shifting holography, in the 1960s in their quest for developing 3D holographic microscopy[41]. Normally, in phase shifting digital holography, 3 step or 4 step phase shifting techniques are used for stability and good quality numerical reconstruction of digital holographic images. But, in the case of quadrature phase-shifting holography, the reconstruction of the original complex object wave without the zero-order and the twin image, requires only two quadrature-phase holograms and two intensity values (namely the object wave intensity and the reference wave intensity). Then, Jung-Ping Liu and Ting-Chung Poon[42] simplified even this two step process and developed a new quadrature phase-shifting digital holographic method where, only two on-axis holograms are required to reconstruct the original

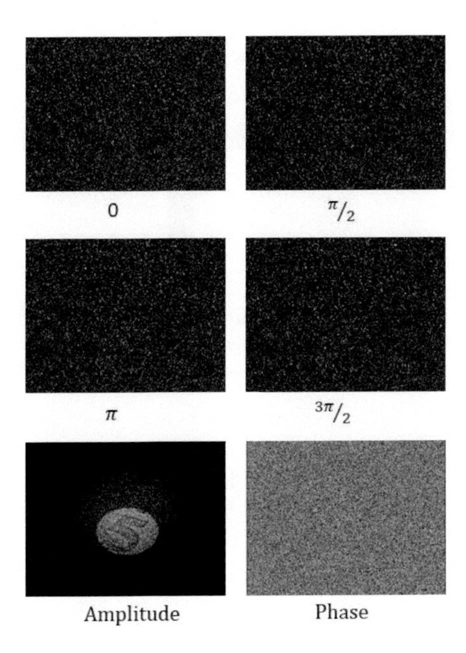

Figure 5.12 Image of 5 Rupees coin recorded using dynamic phase shifting digital holographic set-up shown in Fig. 5.11

complex object wave. This two step quadrature phase shifting digital holographic experimental geometry[43] is shown in Fig. 5.13 where the collimated laser beam passes through a polarizing beam splitter. One of the beams in one state of polarization passes through the object and the other beam acting as reference beam in another polarization state passes through first a half wave plate, then reflecting mirror and a quarter wave plate in a Mach-Zhender interferometric geometry(Fig. 5.13). Both these two beams combine at a non-polarizng cube beam splitter and is recorded by CCD. First hologram of the object is recorded in CCD when the slow axis of quarter wave plate(QWP) is parallel to polarization state of reference beam. Then the second hologram is recorded when the fast axis of quarter wave plate is parrallel to polarization state of reference beam. In this way the QWP will introduce a phase shift of $\frac{\pi}{2}$ between the reference beams in recorded two holograms. Now the intensities recorded after phase shifting of reference beams by 90^0 on CCD array will be,

$$I_{h_1} = |U_r + U_o|^2 = I_0 + 2Re(U_o)U_r \tag{5.68}$$

and after phase shifting the reference beam we get second hologram on CCD arrays as,

$$I_{h_2} = |e^{\frac{j\pi}{2}}U_r + U_o|^2 = |jU_r + U_o|^2 = I_0 + 2Im(U_o)U_r \tag{5.69}$$

where, $I_0 = U_o^2 + U_r^2$ is the zero order intensity of light. If we reconstruct holograms given by Eqns.5.61 and 5.62 then we get twin image noise as well as zeroth order

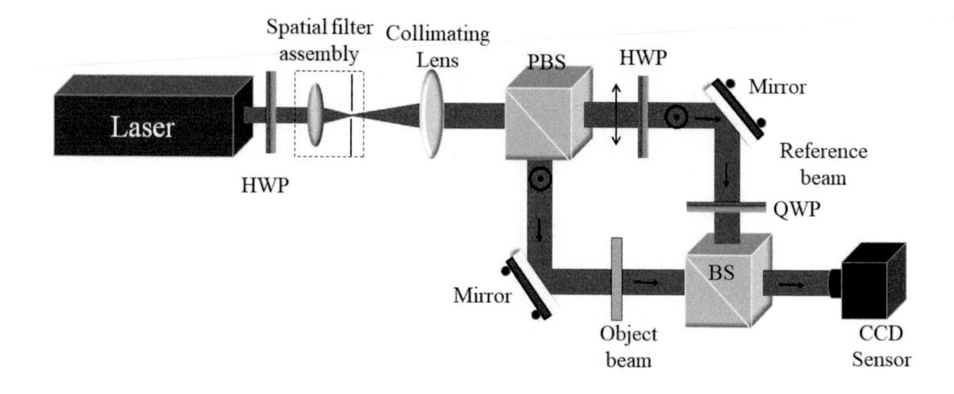

Figure 5.13 Experimental geometry for recording quadratic dynamic phase shifting digital hologram

diffracted light intensity, instead if we combine these two and construct a complex hologram given by,

$$I_{C_{h_1,h_2}} = I_{h_1} + jI_{h_2} = I_o + jI_o + 2I_r[ReU_o + jImU_o] \tag{5.70}$$

$$I_{C_{h_1,h_2}} = I_o + jI_o + 2I_rI_o \tag{5.71}$$

Thus we can get twin image free hologram as $I_{C_{h_1,h_2}}$ is proportional to U_o in addition to zero order light intensity. The zero order light intensity can be completely eliminated if we measure the intensities of I_o, I_r i.e object and reference before holographic recording. Once $I_{h_1} = I_o + I_r$ is found we can obtain twin image free and zero order intensity free digital hologram by subtracting $I_o + jI_o$ from $I_{C_{h_1,h_2}}$, then we have,

$$I_H = (I_{C_{h_1}} - I_o) + j(I_{C_{h_2}} - I_o) \tag{5.72}$$

In this method, we need two quadrature-phase holograms(I_{h_1}, I_{h_2}) and two intensity values i.e I_o, I_r are required to record twin image free and zeroth order intensity free digital holograms. Jung-Ping Liu and Ting-Chung Poon[42] then simplified this 4 recordings further and reduced it to only two steps by using only, two recorded qudrature holograms that is with I_{h_1}, I_{h_2} respectively. Now, by taking square of the absolute value of both sides of Eqn.5.70 we get,

$$2I_o^2 - (4I_r^2 + 2I_{h_1} + 2I_{h_2})I_o + (I_{h_1}^2 + I_{h_2}^2 + 4I_r^4) = 0 \tag{5.73}$$

The solution to above quadratic equation is found to be,

$$I_o = \frac{2I_r^2 + I_{h_1} + I_{h_2}}{2} \pm \frac{\sqrt{(2I_r^2 + I_{h_1} + I_{h_2})^2 - 2(I_{h_1}^2 + I_{h_2}^2 + 4I_r^4)}}{2} \tag{5.74}$$

Thus, we can obtain zero order light, without additional measurements if we know the intensity of reference light. Considering Eqns. 5.68 and 5.69 we get,

$$I_{h_1} + I_{h_2} + 2I_r^2 = 2I_o + 2[I_r + ReU_o + ImU_o]I_r \quad (5.75)$$

Above equation gives,

$$I_o = \frac{2I_r^2 + I_{h_1} + I_{h_2}}{2} - [I_r + ReU_o + ImU_o]I_r \quad (5.76)$$

From Eqns. 5.74 and 5.76 we find that for $I_r + ReU_o + ImU_o = F > 0$ the quantity F will be positive and this requires the value of intensity of reference beam larger which is actually correct in most cases. For lesser value of I_r while reconstruction noises can occur in the reconstructed image. If A is considered as amplitude ratio then,

$$A = \frac{I_r}{\frac{1}{2}[max|U_o| + min|U_o|]} \quad (5.77)$$

and the correlation factor is defined as,

$$CorrelationFactor = \frac{[|E_r|^2 * I_O]_{peak}}{\Sigma |E_r|^2} \quad (5.78)$$

The term $*$ denotes cross correlation operation and I_O is the intensity of original object, Σ represents summation of all data pixels and E_r is the calculated complex amplitude value of object light field at the reconstruction plane at different intensity levels i.e I_r. Thus one can obtain twin image free and zero order diffracted light intensity free digital holographic images using just two steps quadrature recordings by changing referenece beam phase in two steps. Experimental results can be obtained using laser beam of wavelength either 632.8 nm He-Ne laser or 532 nm laser, the object to be recorded can be kept at 5 cm from CCD camera with number of pixels of the object $750(H) \times 750(V)$ with pixel size $3.3\mu m$.

5.5.4 GEOMETRIC PHASE SHIFTING COLOR DIGITAL HOLOGRAPHY

Berry in 1984 [39] showed that the wavefunction of a quantum system can acquire an additional phase factor known as geometric phase when the system is taken around a circuit in parameter space. Later, the geometric phase was measured by Chyba et.al, using a Michelson interferometer geometry [39]. Further, Hariharan et al. showed that the geometric phase ϕ_G shows up in addition to the dynamic phase obtained due to conventional phase shifting method in an interferometric experiment using a Sagnac interferometer[43]. The idea of using geometric phase for digital phase shifting holography for recording color digital holography was introduced by Jun-ichi Kato et.al [44]. This is because for realizing full color phase-shifting digital holography, methods have to be developed to deliver correct phase shifts for each of the red-green-blue (RGB) colors and to record color interferograms simultaneously. Then they have to be reconstructed for unifying the color image with the same

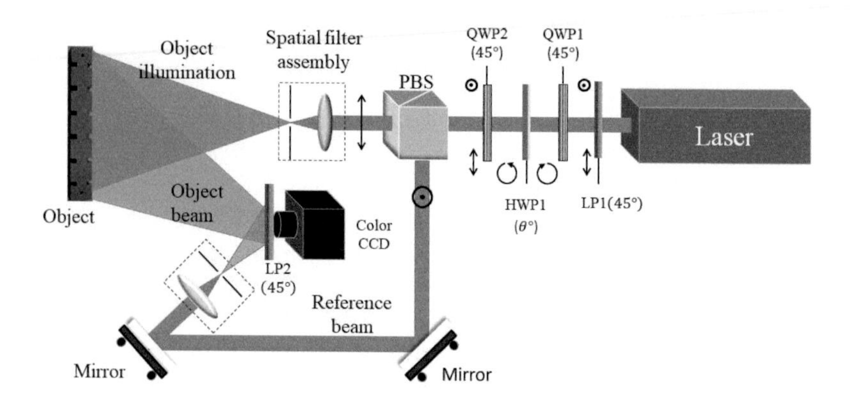

Figure 5.14 Experimental geometry for recording color digital holography using geometric phase shifting.

magnification. For achieving Jun-ichi Kato et.al have developed a quasi-achromatic phase shifter based on the geometric phase. Fig. 5.14 shows the experimental setup used by Jun-ichi Kato et.al for achieving color digital holograms using geometric phase shifting inteferometer geometry where, the laser source used was a He– Cd white-light laser (Koito KLP-450S) with typical wavelengths (powers) of 636.0 nm (9 mW), 537.8 nm (9 mW), and 441.6 nm (27 mW) respectively. These coaxially emitted beams from the respective lasers combine and pass through an achromatic phase shifter based on geometric phase. This achromatic geometric phase shifter consists of a half-wave plate(HWP)sandwiched by a pair of crossed quarter-wave plates, QWP1 and QWP2 respectively as shown in Fig. 5.14. Consider, the laser beam with its orthogonal linear polarization oriented at 45° with respect to the polarizing axis of the laser is converted to left handed circular polarization by the first quarter Wave Plate(QWP1). Then this beam will change to right handed circular polarization after passing through the sandwiched Half Wave Plate(HWP1) as shown in Fig. 5.14. Further this right handed circularly polarized beam after passing through second quarter wave plate(QWP2) become linearly polarized. This process when projected on to a Poincare sphere subtends a solid angle 4ϕ at the center of sphere and gives a phase shift equal to 2ϕ[39]. This phase shift with respect to intial phase of the beam is called geometric phase shift where neither pathlength or wavelength is involved. Now, for getting object and reference beams, a broadband polarization beam splitter, PBS (Newport 10FC16PB.3) was used immediately after the QWP2 as shown in Fig. 5.14. To maintain the intensity ratio of the reference and the object beams, a Neutral Density Filter in the reference beam path is inserted. The object and reference baems are recombined coaxially with a half-mirror, HM, and interfere with another polarizer,PL2, inclined ± 45 from both polarization directions and then reach the CCD plane. The achromatic phase shifter, inserted before the laser beam is expanded to 3-cm diameter, which is required to avoid errors caused by the nonuniform thickness of the mica wave plates. For recording multicolor interferograms, Jun-ichi Kato et.al [44] adopted a digital-color CCD (Flovel AD-21) whose pixel number and pitch are $1636H \times 1236V$ pixels and $3.9mm \times 33.9mm$ respectively. The R(Red)G(Green)B(Blue) images of $818H \times 618V$ pixels are separated with a Bayer color-filter arrangement to generate phase-shifted holograms for each color. While recording the holograms phase differences $\delta\phi$ equalent to $\frac{\pi}{2}$, π and $\frac{3\pi}{2}$ were introduced between the object and the reference waves by rotating the HWP1 by values equal to $\pm 22.5^o$, $\pm 45^o$ and $\pm 67.5^o$ respectively. The rotation angle values of HWP results in $4 \times \phi$ for phase shift between reference and object beams. The complex amplitude of 4 phase shifted color digital holograms is given by,

$$U_h(x,y,\lambda) = [I_h(x,y,\lambda,0) - I_h(x,y,\lambda,\pi)] + i[I_h(x,y,\lambda,\frac{\pi}{2}) - I_h(x,y,\lambda,\frac{3\pi}{2})] \quad (5.79)$$

For recording color digital holography, one has to maintain same focusing distances and mag- nifications at different wavelengths[44]. With the help of Fresnel transformation, and the convolution with two Fast Fourier transforms reconstruction can be obtained. The reconstructed complex amplitude at a distance Z is given by,

$$U_H(X,Y,Z,\lambda) = e^{\frac{2\pi j}{\lambda}\frac{(x^2+y^2)}{2Z}} * U_h(x.y,\lambda) \quad (5.80)$$

In this way, an image area larger than the CCD size $6.4Hmm \times 4.8Vmm$ can be reconstructed by means of embedding the captured holograms in a zero-filled square of 4096×4096 pixels. This process led to a reconstruction area of $32mm^2$ and the entire procedure can be done using a personal computer.

5.5.5 RECONSTRUCTION PROCEDURE

As described in previous section the geometric phase shifting has to be carried out individually for all three RGB wavelengths in which a He–Ne laser (632.8 nm) and an Argon ion laser (514.5 and 457.9 nm) were used. The variations of phases and contrasts against the rotation angle of the HWP1(Half Wave Plate) can be obtained with Fourier analysis of carrier-introduced fringe generated by a tilted plane mirror as an object. There will be difference of the contrast for each color, with low value for the blue line, which is due to the differences in the colors, temporal coherence and the imbalance in the intensity ratio of the reference and the object waves resulting due to the imperfection in achromaticity of the polarizing beam splitter, PBS. In practice, the changes in the contrast or the bias in- tensities among the phase-shifted interferograms lead to a parasitic bias term in the complex amplitude on the CCD. This will generate square noise and is a zeroth-order diffraction in the reconstruction. To overcome these problems, the reference intensities, $I_r(x,y,z,\Delta\phi)$ for each phase shift values of $\Delta\phi$ has to be recorded by blocking the object beam before recording digital hologram and then these intensity values have to be subtracted from the interferograms of both object and reference beams, before final reconstruction. With this correction, the main superimposed noise could be successfully reduced to reconstruct a high contrast color digital hologram[44].

5.5.6 GEOMETRIC PHASE SHIFTING DIGITAL HOLOGRAPY USING MICHELSON INTERFEROMETER GEOMETRY

In the geometric phase shifting technique used by Jun-ichi Kato and Yamaguchi[44] for recording color digital holography, first intensities of the reference beam at different phase shifting values $\Delta\phi$ have to be recorded by blocking object beam. Then these intensities values are subtracted from the combined interferograms. This is because, the laser beam from the source first passes through achromatic geometric phase shifter and then it is expanded and gets split up in to two by a polarizing beam spliiter for object and reference beams(Fig. 5.14). In conventional phase shifting methods only reference beam is phase shifted and in a similar way, Boaz et.al [45] have developed a simple geomteric phase shifting digital holography by modifying Michelson Interferometer used by Chyba et al. [39] to a Twymann-Green interferometer set up. In this only reference beam undergoes geometric phase shifting and not both object and reference beams as in the case described by Jun-ichi kato and Yamaguchi. Consider Fig. 5.15 which, shows the typical experimental geometry based

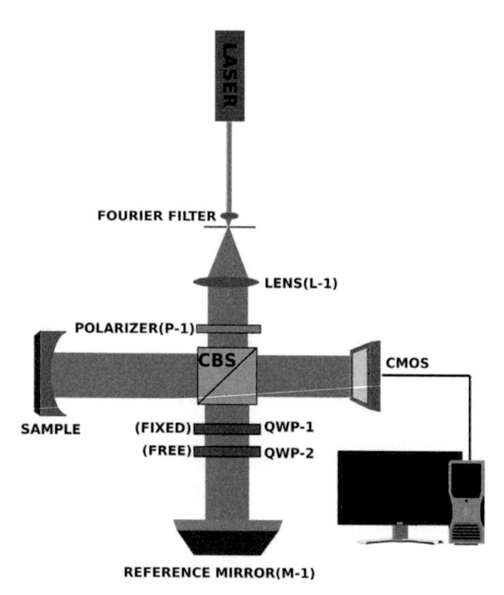

Figure 5.15 Experimental geometry in Twymann Green interferometer set-up.

on Twymann Green Interferometer(Modified Michelson Interferometer). In one of arms(Reference beam) of interferometer two quarter waveplates (QWP_1, QWP_2) are kept and out of thsese two, the first QWP(QWP_1) is fixed in the reference arm so that it can make an angle of 45° with the polarizer(P_1) and the second QWP($QWP2$) is placed behind QWP_1 and is free to rotate. In the other arm of the experimental set up an object is kept as shown in Fig. 5.15. The laser beam from a He-Ne laser is spatially filtered and collimated using lens(L_1). This collimated beam, becomes linearly polarized after it passes through a polarizer(P_1). This linearly polarized beam is divided equally into both the arms by a non-polarizing cube beam splitter(CBS). One of the beams from the cube beamsplitter directly illuminates the object and the diffusely reflected beam from the object, becomes object beam and it retraces its path to reach back beamsplitter. The other beam is the reference beam which passes through two quarter wave plates and retraces its path via the two quarterwave plates after getting reflected from mirror(M2) to reach the beam spliiter as shown in Fig. 5.15. These two beams interfere at the CCD or CMOS plane to record digital hologram. The geometric phase shifting process in this geometry follows original experimental setup of Chyba et.al[39] in a Michelson Intereferometer geometry and can be explained using a Poincare sphere. In this geometric phase shifting process the reference beam only undergoes phase shifting with respect to objcet beam and therefore, If E_{x_r} and E_{y_r} represent the x and y components of reference beam and are mutually perpendicular and transverse to the direction of propagation then using Wolf's coherency matrix calculus for 2 poalrzation matrix is given by,

$$J_{c_r} = \begin{bmatrix} < E_{x_r}E^*_{x_r} > & < E_{x_r}E^*_{y_r} > \\ < E_{y_r}E^*_{x_r} > & < E_{y_r}E^*_{y_r} > \end{bmatrix} = \begin{bmatrix} J_{x_r,x_r} & J_{x_r,y_r} \\ J_{y_r,x_r} & J_{y_r,y_r} \end{bmatrix} \tag{5.81}$$

The Eqn.5.81 is the coherency matrix for the reference beam. This is because we have inserted two quraterwave plates along the path of reference beam and it is necessary to introduce coherency matrix concept for geometric phase shifting. Introducing stoke's parameters to the matrix elements shown in Fig. 5.81 we get,

$$J_{c_r} = \begin{bmatrix} < E_{x_r}E^*_{x_r} > & < E_{x_r}E^*_{y_r} > \\ < E_{y_r}E^*_{x_r} > & < E_{y_r}E^*_{y_r} > \end{bmatrix} = \frac{1}{2}\begin{bmatrix} (S_0 + S_1) & (S_2 + iS_3) \\ (S_2 - iS_3) & (S_0 - S_1) \end{bmatrix} \tag{5.82}$$

now the coherency matrix J_{c_r} can be expanded as linear superposition of the Stokes parameters as,

$$J_{c_r} = \frac{1}{2}\Sigma_{i=0}^{3}\sigma_i S_i \tag{5.83}$$

where σ_i represents Pauli's spin matrices and the polarization states of light beam can be represented in terms of coherency matrix as,

$$J_{LHP} = \frac{1}{2}\begin{bmatrix} 1 & 0 \\ 0 & 0 \end{bmatrix}, J_{LVP} = \frac{1}{2}\begin{bmatrix} 0 & 0 \\ 0 & 1 \end{bmatrix}, \tag{5.84}$$

and for the right and left circular polarization we get,

$$J_{RCP} = \frac{1}{2}\begin{bmatrix} 1 & i \\ -i & 1 \end{bmatrix}, J_{LCP} = \frac{1}{2}\begin{bmatrix} 1 & -i \\ +i & 1 \end{bmatrix} \tag{5.85}$$

Now, the intensity pattern of the reference beam can be written as,

$$I_r(\theta,\phi) = J_{x_rx_r}cos^2\theta + J_{y_ry_r}sin^2\theta + J_{x_ry_r}e^{-i\phi}cos\theta sin\theta + J_{y_rx_r}e^{i\phi}sin\theta cos\theta \tag{5.86}$$

In above Eqn.5.86 the values of $J_{x_rx_r} = E_{x_r}E^*_{xr}$, $J_{x_ry_r} = E_{x_r}E^*_{yr}$, $J_{y_rx_r} = E_{y_r}E^*_{xr}$, $J_{y_ry_r} = E_{y_r}E^*_{yr}$ respectively and with all above explanations the coherency matrix of reference beam can be written as,

$$J_{c_r} = \begin{bmatrix} (1+cos2\theta) & e^{-i\phi}sin2\theta \\ e^{i\phi}sin2\theta & (1-cos2\theta) \end{bmatrix} \tag{5.87}$$

In Eqn 5.87, the θ and ϕ terms can be represented on a Poincare sphere(Fig 5.16) as polar and azimuthan angle respectively in spherical polar co-ordinates, when the linearly polarized reference beam traverses through a static and rotating quarterwave plates for achieving geometrical phase shift. The entire process of geometrical phase shifting can be explained using the Poincare sphere shown in Fig. 5.16. Consider the

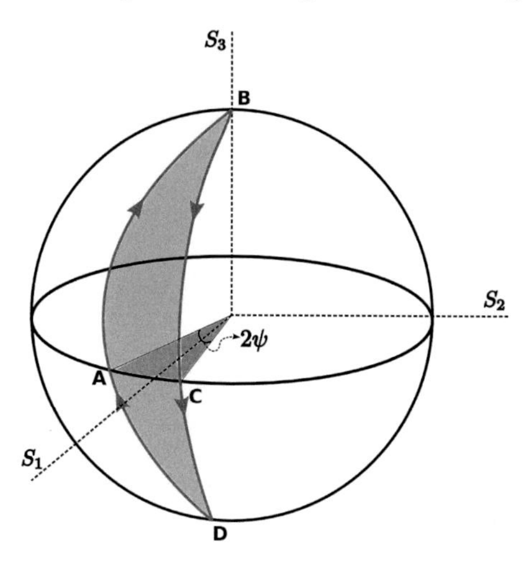

Figure 5.16 Poincare sphere depicting polarization states of reference beam passing through quarter wave ($\frac{\lambda}{4}$) plates Q_1, Q_2.

linearly polarized reference beam leaving the beamsplitter be represented by point A in the Poincaré sphere. This beam after passing through QWP_1 , becomes right circularly polarized light, which is represented by point B in Fig. 5.16. This right circularly polarized reference beam then passes through QWP_2 , which is free to rotate through any angle by ψ. Now, because of passing through the second quarter wave plateQWP_2 this beam becomes linearly polarized and reaches the point C on the equator of Poincare Sphere, which also depends on the rotation angle ψ. The linearly poalrized beam is now reflected by the mirror M_2, and it retraces its path through quarter wave plates QWP_2,QWP_1 respectively and becomes linearly polarized again and reaches back to point A on the Poincare sphere there by, forming a closed loop, as shown in Fig. 5.16. This closed loop motion of light beam subtends a solid angle on the Poincaré sphere and is equal to $(\phi_2 - \phi_1)(cos\theta_1 - cos\theta_2)$. The rotation of QWP_2 by an angle of ψ corresponds to the geodesic AC in the Poincaré sphere given by $\Delta\phi = (\phi_2 - \phi_1)$, and the value of $\psi = 2\Delta\phi$ which results in a solid angle $4\Delta\phi = 2\Delta\psi$ on the Poincaré sphere. This makes the reference beam suffer a geometric phase shift equalent to $2\Delta\psi$ when the quarter wave plate QWP_2 is rotated through an angle of

ψ. This phase 2ψ is the geometric phase ϕ_G, which does not depend on the optical path length or wave- length, unlike the dynamic phase ϕ_D. Thus the phase of the reference beam suffers a pure geometric phase shift equal to 2ψ with respect to the object beam. This process of geometric phase shifting can be explained using Jone's matrix as follows,

$$\begin{bmatrix} -1 & 0 \\ 0 & -1 \end{bmatrix}\begin{bmatrix} 1 & 0 \\ 0 & +i \end{bmatrix}\begin{bmatrix} -1 & 0 \\ 0 & -1 \end{bmatrix}\begin{bmatrix} 1 & 0 \\ 0 & -i \end{bmatrix}\frac{1}{\sqrt{2}}\begin{bmatrix} 1 \\ 1 \end{bmatrix} = \frac{1}{\sqrt{2}}\begin{bmatrix} 1 \\ 1 \end{bmatrix} \tag{5.88}$$

In obtaining Eqn.5.88, we have assumed unit amplitude of reference beam and a rotation matrix with the slow axis of first quarter wave plateQWP_1 being horizontal to the vertical slow axis of freely rotating second quarter wave plateQWP_2 respectively. Now, considering $E_o e^{-\delta_o}$ and $E_r e^{-\delta_r}$ represent the complex ampltudes of object and reference beams then the intensity pattern of hologram at the CCD plane will become,

$$I_h = |E_o e^{-\delta_o} + E_r e^{-\delta_r + 2\psi}|^2 \tag{5.89}$$

$$I_h = I_o + I_r + 2\sqrt{I_o I_r cos(\delta_{or} - 2\psi)} \tag{5.90}$$

where, $\delta_{or} = (\delta_o = \delta_r)$ and the Equation 5.90 represents the intensity of the hologram recorded on a CCD/CMOS camera. The phase shifting process to record digtla hologram can be obtained by rotating the QWP_2 through an angle $\psi = 0; \frac{\pi}{4}; \frac{\pi}{2}; \frac{3\pi}{4}$ respectively, then we can generate phase shifts of $2\psi = 0^o; 90^o; 180^o; 270^o$ respectively. This gives rise to four phase shifted holograms ($I_1; I_2; I_3$, and I_4) which can be used in the four bucket method to generate the complex amplitude in the detector plane similar to dynamic phase shifting hologram. Further, by computing the backpropagation of the complex amplitude using the diffraction formulas , the complex amplitude at the object plane can be obtained. From the complex amplitude information the irradiance and surface profile of the object can be reconstructed.

5.5.7 EXPERIMENTAL RESULTS AND DISCUSSION

The experimental geometry to obtain geometric phase shifting digital hologram can be achieved using the same Twymann Green Interferometer geometry which is a modified Michelson Interferometer. Fig. 5.15 is nothing but Twymann Green geometry in which one can use simple He-Ne laser of minimum power. To demonstrate the capability of geometric phase shifting digital holography, Boaz et.al[45] have used a He–Ne laser emitting at 632.5 nm with 8 mW is used as the light source. The detector used in that experiment was a Motion Pro Y4 series camera, with 13.7 μ m pixel pitch, configured to acquire 768 × 768 sized frames each at an exposure time of 33 ms. The object used in this experiment were two dices as objects kept at "dice-1" and "dice-2" [Fig. 5.17], were chosen and placed at a distances 81 and 91 cm, respectively, from the detector in one of arms of Twyman Green interferometer set up as shown in Fig. 5.15 The actual distances of the dices were measured with respect to the face of the dice facing the CCD camera. The face of dice-1 inscribed with the number 1 faces the camera, while dice-2 had its face inscribed with the number 5 facing the camera. Each cubic die is made up of metla and had side lengths

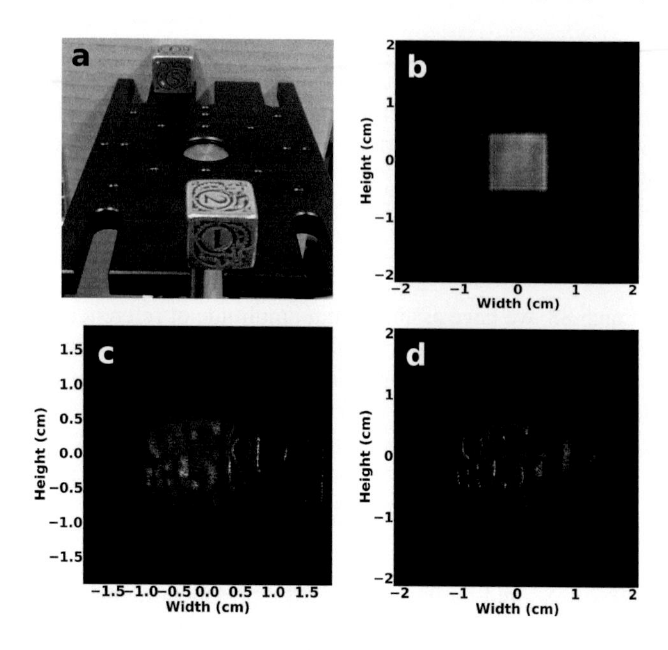

Figure 5.17 Digital holographic image recorded using geometric phase shifting digital holographic set-up shown in Fig 5.15

of 1.5 cm. Since the object, dice reflects less light com pared to the reference mirror, neutral density filters were added in the reference arm to equate intensity for increasing the fringe contrast. Similar to dynamic phase shifting digital holography, four holograms were recorded one after the other by rotating the QWP-2 in steps of angle $\frac{\pi}{4}$. Then the complex amplitude of the wave front at the detector plane E_h was cal- culated from the four resulting holograms ($I_1;I_2;I_3$ and I_4) using the following relation as,

$$E_h = (I_4 - I_2) - i(I_1 - I_3) \tag{5.91}$$

Using Fresnel diffraction formula E_h was back propagated up to distances 81 and 91 cm that corresponded to the origi- nal positions of objects, dice-1 and dice-2, respectively. Intensity was then calculated from the backpropagated complex amplitude, and the results are shown in Figs. 5.17(c) and 5.17(d), respectively. In Fig. 5.17 the face of dice-1 with number 1 is in focus whereas in Fig. 5.17 the face with number 5 is in focus, which corre- sponds to dice-2. The Fresnel back progagation algorithm results in a change in lateral magnification with propagation distance, and hence scales in Figs. 5.17 c and 5.17 d are different. Moreover, the width of both dice put together constitutes 3 cm, while the width of the complementary metal oxide semiconductor (CMOS) sensor is only 1 cm. The demagnifying property of the Fresnel algorithm is the one that made the reconstructions successful in spite of the size differences. This is also the reason behind choosing the Fresnel diffraction formula for the recon- structions in this experiment. The right side edge of dice-2 is not visible from the reconstruction shown in Fig. 5.17. This is due to the fact that a small portion of the inner edge of dice-2 was hidden from the illumination by the overlapping dice-1 (due to smaller beam size). Figure 5.17(b) shows the reconstruction of dice-2 again, but by using only the intensity I_4 for backpropa- gation. It is clearly seen that the dc component is very strong and masks the necessary object information during reconstruction.Thus it is clear that the geometric phase shifting can be used for 3D holographic imaging applications similar to dynamic phase shifting holography.

6 Unconventional holography

6.1 COHERENCE HOLOGRAPHY

6.1.1 INTRODUCTION

We have discussed so far about several methods using the principle of conventional holography where it records and reconstructs the 3-D image of an object represented by an optical field distribution itself. In 2005, Mitsuo Takeda et.al[46] has proposed and demonstrated a new type of holography, called coherence holography. In coherence holographic technique, the information of the 3-D image of the object is encoded into the spatial coherence function of the reconstructed optical field. Unlike conventional holography, this new type of coherence holography reconstructs the image as the degree of spatial coherence between a pair of points, of which one serves as a reference point R and the other as a probe point P on the object to be reconstructed.

6.1.2 PRINCIPLE OF COHERENCE HOLOGRAPHY

The principle of coherence holography is based on the formal analogy between the diffraction integral and the formula of Van Cittert-Zernike theorem [Appendix D]. To explain the basic principle of coherence holography, let us consider the case of phase-conjugate reconstruction in conventional holography, in which the illuminating light for recording and reconstruction is normally drawn from laser sources which are temporally and spatially coherent. Fig. 6.1 shows a typical reconstruction geometry using a phase conjugate beam in conventional holography where, initially a three-dimensional object is recorded in a hologram with a spherical reference beam $E_R(\mathbf{r_s}) = \frac{e^{ik(|\mathbf{r_s}-\mathbf{r_R}|)}}{|(\mathbf{r_s}-\mathbf{r_R})|}$ diverging from a point source R. Then the holographic image is reconstructed by illuminating the hologram with the phase-conjugated reference beam $E_R^*(\mathbf{r_s}) = \frac{e^{-ik(|\mathbf{r_s}-\mathbf{r_R}|)}}{|(\mathbf{r_s}-\mathbf{r_R})|}$ which is also a converging beam into the original reference source point R, as shown in Fig. 6.1. In that, $\mathbf{r_s}$ and $\mathbf{r_R}$ are position vectors pointing at point S on the hologram and the reference point source R, respectively. Now, the reconstructed optical field at an arbitrary observation point Q is given by,

$$
\begin{aligned}
E_h(\mathbf{r_Q}, \mathbf{r_R}) &= \int\int A_h(\mathbf{r_s}) E_R^*(\mathbf{r_s}) \frac{e^{ik(|\mathbf{r_Q}-\mathbf{r_s}|)}}{|(\mathbf{r_Q}-\mathbf{r_s})|} d\mathbf{r_s} \\
&= \int\int A_h(\mathbf{r_s}) \frac{e^{ik(|\mathbf{r_Q}-\mathbf{r_s}|-|\mathbf{r_R}-\mathbf{r_s}|)}}{|(\mathbf{r_Q}-\mathbf{r_s})||(\mathbf{r_R}-\mathbf{r_s})|} d\mathbf{r_s} \qquad (6.1)
\end{aligned}
$$

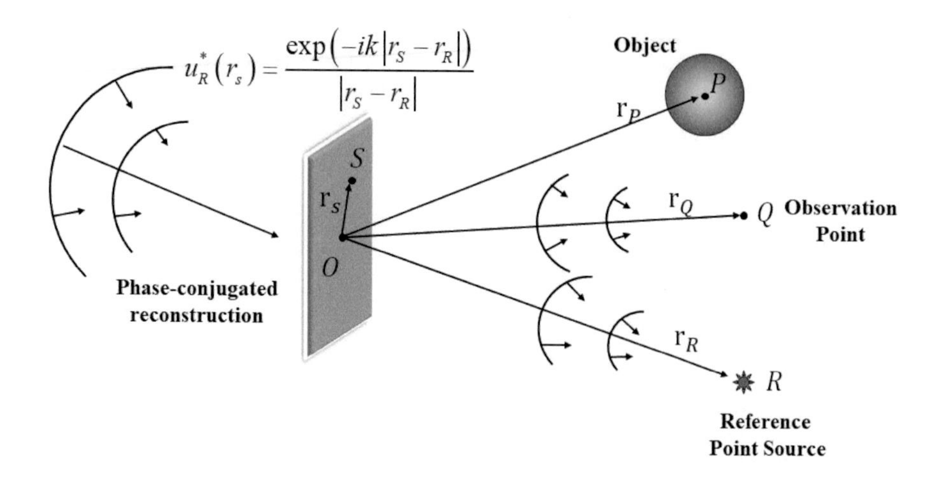

Figure 6.1 Reconstruction of an optical field from a conventional hologram using phase cojugate beam.

where, $A_h(\mathbf{r_s})$ represents amplitude transmittance of the hologram and $\mathbf{r_Q}$ is the position vector of the observation point Q with the integration taken over entire hologram. It can be seen that the Eqn 6.1 resembles similar to the mutual intensity formula given by the Van Cittert–Zernike theorem [Appendix D]as,

$$J(\mathbf{r_Q}, \mathbf{r_s}) = \int\int A_l(\mathbf{r_s}) \frac{e^{ik(|\mathbf{r_Q}-\mathbf{r_s}|-|\mathbf{r_R}-\mathbf{r_s}|)}}{|(\mathbf{r_Q}-\mathbf{r_s})||(\mathbf{r_R}-\mathbf{r_s})|} d\mathbf{r_s} \tag{6.2}$$

where, $I_l(\mathbf{r_s})$ is the intensity distribution of the spatially incoherent source. Now the analogy between the formula for mutual intensity and that for the complex optical field reconstructed from a hologram with a phase-conjugated beam can be explained as follows. Consider Fig. 6.1 where the intensity transmittance of hologram is given by $I_l(\mathbf{r_s}$ and if this intensity is proportional to the recorded intensity, illuminated with quasi-monochromatic spatially incoherent light with a temporal coherence length sufficiently larger than the longitudinal depth of the 3-D object, then an optical field can be generated. For this, the mutual intensity between observation point Q and the reference point R is equal to the optical field that would be reconstructed if the hologram with the same amplitude transmittance $A_h(\mathbf{r_s}) = A_l(\mathbf{r_s})$ were illuminated with a phase-conjugated version of the reference beam. Unlike conventional holography, the reconstructed coherence image of the coherence holography is not directly observable and it can be visualized only as the contrast and the phase of an interference fringe pattern by using an appropriate interferometer. Because of this reason this technique of holography is called as coherence holography. This is similar to a conventional computer-generated hologram (CGH)(Chapter 3) which can create a three-dimensional image of a non-existing object and the computer-generated coherence hologram (CGCH) can create an optical field with a desired three-dimensional distribution of a spatial coherence function based objects. If the hologram is illuminated with a partially coherent reference light with complex degree of coherence function ($\mu(\mathbf{r_s}, \mathbf{r_r})$ then corresponding mutual intensity of reconstructed field can be written as,

$$J(\mathbf{r_Q}, \mathbf{r_R}) = \int\int\int\int J(\mathbf{r_s}, \mathbf{r_s}) \frac{e^{ik(|\mathbf{r_Q}-\mathbf{r_s}|-|\mathbf{r_R}-\mathbf{r_s}|)}}{|(\mathbf{r_Q}-\mathbf{r_s})||(\mathbf{r_R}-\mathbf{r_s})|} d\mathbf{r_s} d\mathbf{r_s} \tag{6.3}$$

where, $J(\mathbf{r_s}, \mathbf{r_s}) = \sqrt{I_l(\mathbf{r_s})}\sqrt{I_l(\mathbf{r_s})}\mu(\mathbf{r_s}, \mathbf{r_r})$ is the nutual intensity function immediately behind the hologram with $I_l(\mathbf{r_s})$, as the intensity transmittance of the hologram. Suppose, the spatialcoherence is stationary and the field correlation legth is shorter than the spatial structure of the hologram, then we can write $J(\mathbf{r_s}, \mathbf{r_s}) = I_l(\mathbf{r_s})\mu(\Delta\mathbf{r_s})$ with $\Delta\mathbf{r_s} = \mathbf{r_s} - \mathbf{r_s}$. Further, by changing the variable of integration of Eqn.6.3, we get,

$$J(\mathbf{r_Q}, \mathbf{r_R}) = \int\int \mu(\Delta\mathbf{r_s})[\int\int I_l(\mathbf{r_s}) \frac{e^{ik(|\mathbf{r_Q}-\mathbf{r_s}|-|(\mathbf{r_R}-\Delta\mathbf{r_s})-\mathbf{r_s}|}}{|(\mathbf{r_Q}-\mathbf{r_s})||(\mathbf{r_R}-\Delta\mathbf{r_s})-\mathbf{r_s}|}]d(\Delta\mathbf{r_s}) \tag{6.4}$$

Since, $A_h(\mathbf{r_s}) = A_l(\mathbf{r_s})$ and substituting that in Eqn.6.1, we get,

$$J(\mathbf{r_Q}, \mathbf{r_R}) = \int\int \mu(\Delta\mathbf{r_s}) E_H(\mathbf{r_Q}, \mathbf{r_R} - \Delta\mathbf{r_s}) d\Delta\mathbf{r_s} \tag{6.5}$$

The Eqn.6.5 states that if the illumination is partially coherent, the coherence function reconstructed from the coherence hologram will be blurred, and will be convolved with function $(\Delta\mathbf{r_s})$ of the complex degree of coherence function. On the other hand, if the illumination is perfectly spatially incoherent, then we get $\mu(\Delta\mathbf{r_s})$ proportional to $\delta(\mathbf{r_s})$, which results in $J(\mathbf{r_Q}, \mathbf{r_R}])$ proportional to $E_h(\mathbf{r_Q}, \mathbf{r_R})$, so that recorded 3-D objectfrom a hologram can be ideally reconstructed as the distribution of the mutual intensity. Thus the relation between conventional holographic methods and coherence holographic method lies in reconstruction using a spatially incoherent light. In case of conventional incoherent holography,the reconstruction of the object is done physically by illuminating the hologram with coherent light, while in case of the generalized Michelson stellar interferometer[46], the reconstruction of the object is done numerically from the spatial coherence function detected with an interferometer. In case of coherence holography, however, the hologram is recorded with coherent light and is reconstructed with spatially incoherent light. In this sense, the role of coherent and incoherent light is reversed between coherence holography and conventional incoherent holography. Also, in case of the generalized stellar interferometer, it records a coherence function and reconstructs an intensity distribution, but in case of coherence holography, it records the fringe intensity and reconstructs the mutual coherence function. This further shows that, the role of the intensity and the coherence function is also reversed between coherence holography and the stellar interferometer based incoherence holography.

6.1.3 EXPERIMENTAL PROCEDURE

Since in case of conventional holographic method which was just described in previous section, the reconstructed image encoded in the spatial coherence of the field cannot be observed directly, one could probe the coherence of the field with a Young's double slit experimental setup[Appendix A]. Young's double slit procedure is a point-probing scheme, which requires one to scan one of the slits over the 3-D space to probe the coherence of the field. To simultaneously visualize the full-field distribution of the spatial coherence function as the contrast and the phase distribution of interference fringes, an appropriate interferometer such as a wave-front-folding interferometer [46] and a triangular interferometer with a double afocal system [46] can be considered. Fig. 6.2 shows a Fizeau interferometer, where, the two

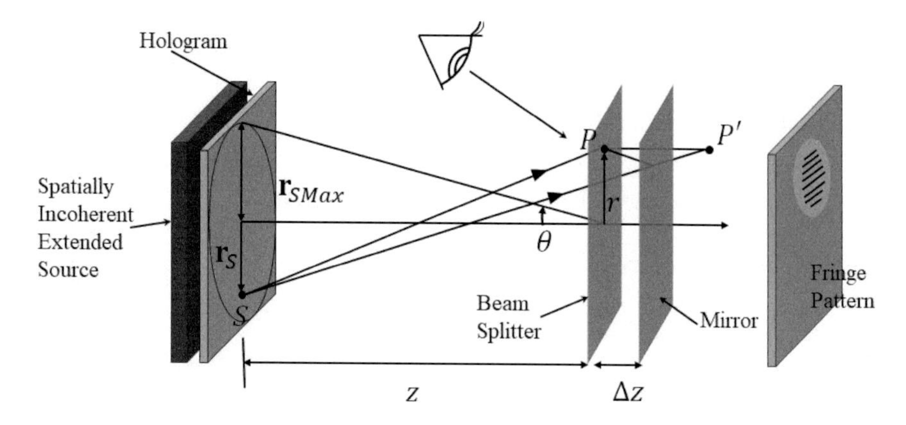

Figure 6.2 Visualization of a coherence image reconstructed from a coherence hologram.

plane parallel plates formed by the beam splitter and the mirror produces two point sources resulting in an interference pattern similar to Fresnel zone plate equivalent to $I(\mathbf{r_S})$. Because of this the intensity based superposition of these Fresnel Zone plates weighed by the intensity of hologram, we can observe interference fringe intensity at point O as a complex degree of coherence and can be written as,

$$I(\mathbf{r},dz) = [\int I_s(\mathbf{r_s})d\mathbf{r_s}][1+|\Gamma(\mathbf{r},dz)|cos[\phi(dz)-\psi(\mathbf{r},dz)] \tag{6.6}$$

In above Eqn $\phi(dz)$ is initial phase of Fresnel Zone plate fringe and $\Gamma(\mathbf{r},dz)$ is a complex degree of coherence function and is equal to,

$$\Gamma(\mathbf{r},dz) = \frac{\int I_s(\mathbf{r_s})e^{\frac{ikdz|\mathbf{r}-\mathbf{r_s}|^2}{z^2}})d\mathbf{r_s}}{\int I_s(\mathbf{r_s})d\mathbf{r_s}} \tag{6.7}$$

The complex degree of coherence factor $\Gamma(\mathbf{r},dz)$ is actually the Fresnel transform of incoherently illuminated hologram. Thus if a Fresnel hologram with coherent light of an object at distance $d = \frac{z^2}{2dz}$ is recorded and illuminated it with an incoherent light from behind, then we can observe a set of interference fringes on the beam splitter representing field amplitude and phase of the original object which was recorded with a coherent light source.

6.1.4 EXPERIMENTAL DEMONSTRATION OF COHERENCE HOLOGRAPHY

In this experiment, a modified Sagnac radial shearing interferometer with a projector illumination as phase shifter is employed together with phase-shift fringe analysis. A simple implementation of off-axis coherence holography with a commercial projector combined with a Sagnac radial shearing interferometer is shown in Fig. 6.3 The axial shear accompanying the radial shear comes handy in the 3-D object reconstruction without affecting the temporal coherence between the interfering beams in a common path Sagnac interferometer. As the Sagnac common path interferometer is very stable to environmental noises caused by vibrations and air turbulences, the reconstruction process is highly reliable. A set of phase-shifted Fourier transform holograms was displayed sequentially with the projector. The coherence function was detected from the corresponding interferograms and the object was reconstructed as the 3-D correlation map of the fields diffracted from the hologram. The implementation of 3-D coherence holography using only a commercially available projector for the display of hologram truly met the requirement that only an incoherent light source is needed to reconstruct a hologram in coherence holography. The coherence holo-

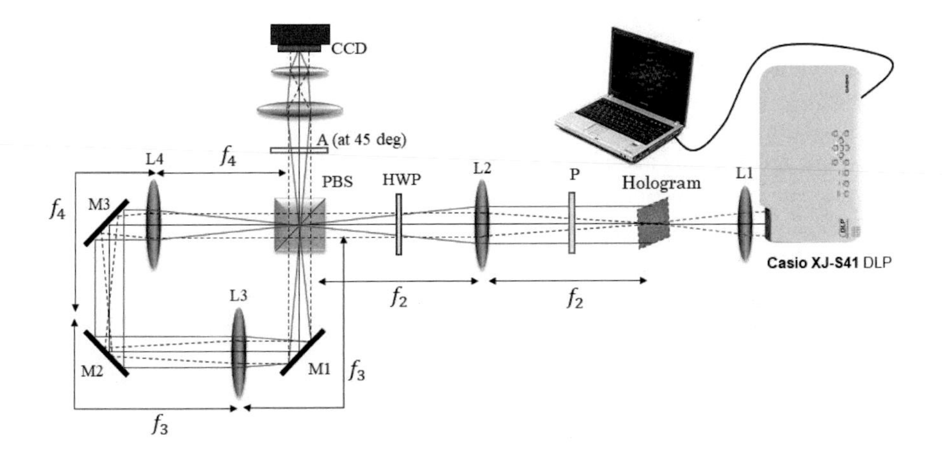

Figure 6.3 Experimental set-up for demonstration of coherence holography.

grams used in the experiment is a set of numerically generated phase shifted Fourier transform holograms. The alphabets C and H as shown in Fig 6.4(a) and 6.4(b) with sizes about 80 × 80 pixels each kept at different depth locations were used as the objects.The Fourier transform of these off-axis binary objects are used to generate the coherence hologram. Fig. 6.4(c) shows one of the phase shifted coherence holograms used in the experiment. The experimental set up is shown in Fig. 6.3. The first

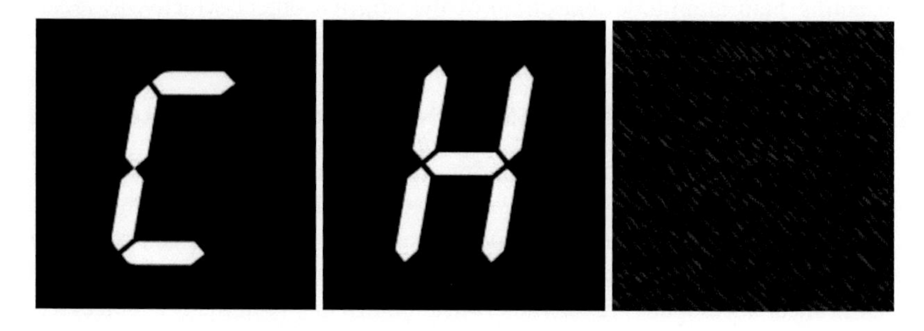

Figure 6.4 (a) and (b) are letters C and H used as objects in the experiment and (c) is one of the phase shifted coherence holograms used in the experiment.

part of the experimental set-up is meant for the display of incoherently illuminated coherence hologram which is implemented by a commercial projector having a resolution of 1280×1024 pixels. The magnification of the projected hologram is suitably controlled with the proper choice of Lens L_1. The hologram is imaged on the front focal plane of Lens L_2.The field distribution of the incoherently illuminated hologram is Fourier transformed by lensL_2 with a focal length 100*mm*. A beam splitter (PBS) splits the incoming beam into two counter propagating beams. The telescopic system with magnification 1.04 , formed by lenses L_3 (focal length 125*mm*) and L_4 (focal length 120*mm*), gives a radial and axial shear between the counter propagating beams as they travel through interferometer before they are brought back together and imaged by CCD. The resulting interference gives a 3-D field correlation distribution that reconstructs the image as a coherence function represented by the fringe contrast. The coherence function was detected by applying the phase-shift technique to the Sagnac radial shearing interferometer, and the object was reconstructed as the

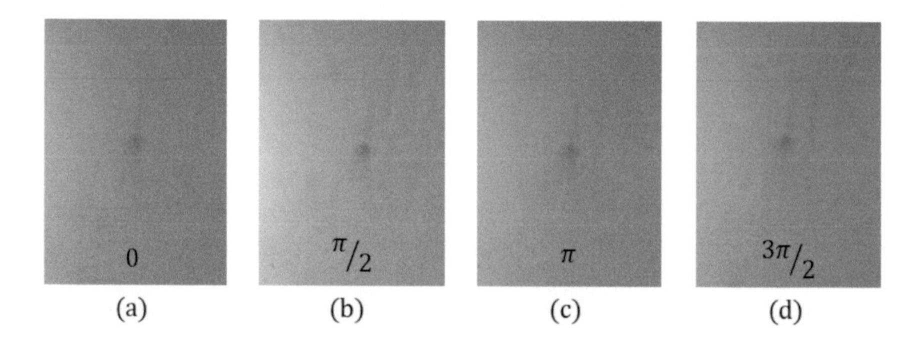

Figure 6.5 The phase shifted interference fringes recorded with CCD in the experiment.

3-D correlation map of the fields of diffracted from the hologram. At the output of the interferometer, by scanning the CCD camera along z direction the corresponding part of the reconstructed object gets imaged on the CCD image plane. At each location of the CCD camera, the set of phase- shifted Fourier transform holograms was displayed sequentially with the projector and the corresponding interferograms were captured by a CCD camera. Figures 6.5 shows the 4 phase shifted interferograms recorded at CCD plane. Figs 6.6(a) shows the corresponding fringe contrast, and Figures 6.6(b) shows, the corresponding fringe phase calculated using same 4 step phase shift reconstruction method. Thus one can conveniently use coherence holography using incoherent sources for reconstruction.

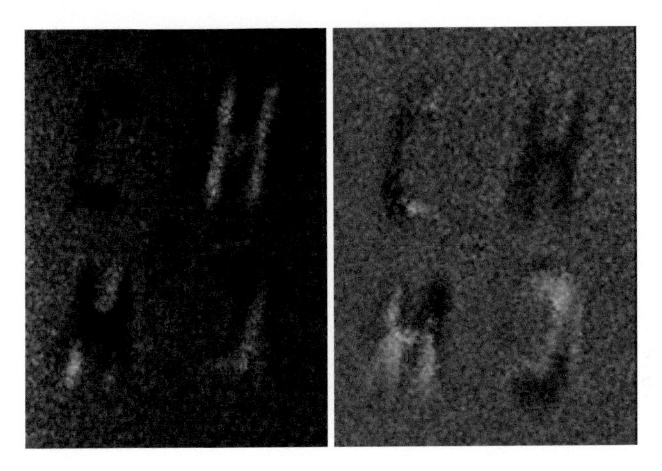

Figure 6.6 (a) shows the fringe visibility; and (b) shows the fringe phase.

A Coherence of Optical waves

A.1 INTRODUCTION

The coherence of optical beams play very important role in obtaining interference pattern of two light beams and more importantly for getting good quality holograms. In general, the coherence property is directly linked to the origin of light beams from the source. For example, in case of laser beam generation, the life time of atoms in meta stable state is direcly related to coherence length of laser beam. Consider a wavefront with its electric field component given by $E = E_0 sin(kz - \omega t + \phi)$ and after some time i.e at $(t = t + \Delta t)$ if this wavefront still remains sinusoidal with same phase then these two wavefronts are coherent. Thus at time t and $(t + \Delta t)$ if two wavefronts have same phase relationships then $\Delta t << \tau_c$ and if they do not have same phase relationship then $\Delta t >> \tau_c$ where τ_c is called coherence time and is linked with coherence length as $l_c = c\tau_c$ with c is velocity of light and l_c is coherence length respectively. At any two positions in space that is at $z = z_1$, and at $z = z_2$ if the wave $E_1 = E_{0_1} sin(kz_1 - \omega t + \phi)$, and $E_2 = E_{0_2} sin(kz_2 - \omega t + \phi)$ have same phase relationships then these two beams E_1, E_2 are spatially coherent. On the other hand if at any instant of time i.e $E_1 = E_{0_1} sin(kz_1 - \omega t_1 + \phi)$, and $E_2 = E_{0_2} sin(kz_2 - \omega t_2 + \phi)$ have same phase relationship then these two beams are said to be temporarily coherent.

A.2 SPATIAL COHERENCE

The spatial coherence phenomena of light can be explained using Young's double slit experiment. In practice all practical light sources have finite physical sizes. To understand spatial coherence phenomena of light source consider two light signals $U(s_1,t)$, $U(s_2,t)$ are being observed at two points in space s_1, s_2 with no time delay. If $s_1 = s_2$ then the two wavefronts $U(s_1,t)$, $U(s_2,t)$ are perfectly correlated but if they are separated laterally in space then graduoly the correlation between these two points fall. This shows that the two waves emitted from the source have limited spatial coherence. This can be explained using Young's double slit experiment. Consider Fig. A.1 in which an Young's double slit experimental geometry is shown. In that from an extended light source, two slits S_1, S_2 are kept at a distance l. Also, a screen is kept at a distance z behind the slits. If the two slits are sufficiently close then good contrast interference fringes can be seen on the screen when the light beam from source S illuminates the slits. On the other hand if the distance between these two slits d is large then contrast of interference fringes seen on the screen falls. This spatial coherence phenomena can be understood by changing the distance between

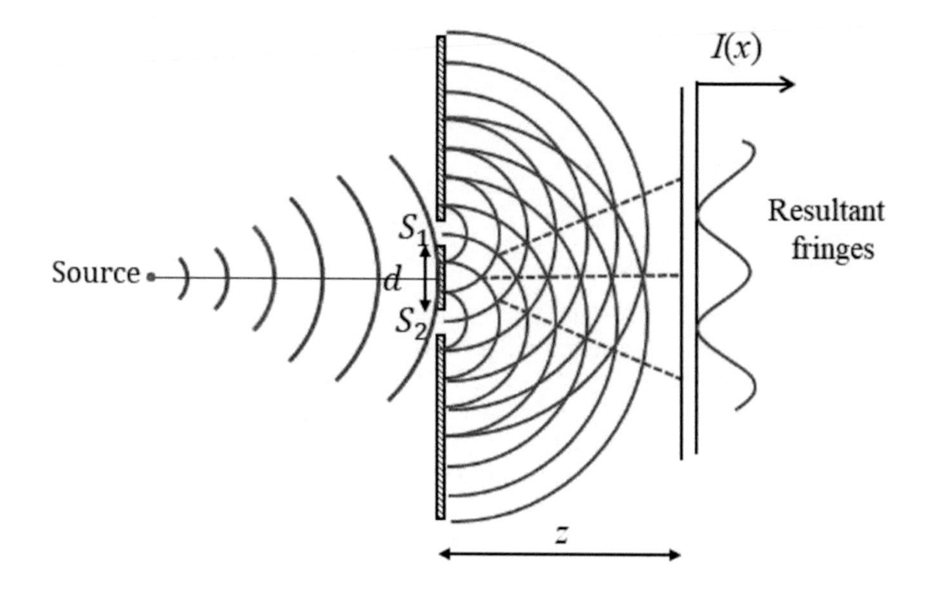

Figure A.1 Young's double slit experimental geometry for demonstrating spatial coherence of light beam.

slits i.e d from the Young's double slit experiment. Let us consider O the point on the screen and the light beam from slit S_1, S_2 reaches it with time delays $(\frac{r_1}{c}, \frac{r_2}{c})$ respectively. Now, if the time delay between these two beams is given by $\frac{(r_2-r_1)}{c}$ is less than coherence time of source i.e τ_c then high contrast interference can be obtained and this can be experimentally observed by moving the distance between two slits d. Fig. A.2 shows the interference pattern when two slits have smaller separation of slits and Fig. A.3 shows the interference fringes when two slits S_1, S_2 have larger separation. One can see the loss of contrast or visbility of interference fringes in Fig. A.3. This is due to low spatial coherence.

A.2.1 THEORETICAL EXPLANATION

The intensity of light beams at point O on the screen in Fig. A.1, is given by,

$$I(O) = < U(O,t)U^*(O,t) > \quad \text{(A.1)}$$

where,

$$U(O,t) = C_1 U(s_1, t - \frac{r_1}{c}) + C_2 U(s_2, t - \frac{r_2}{c}) \quad \text{(A.2)}$$

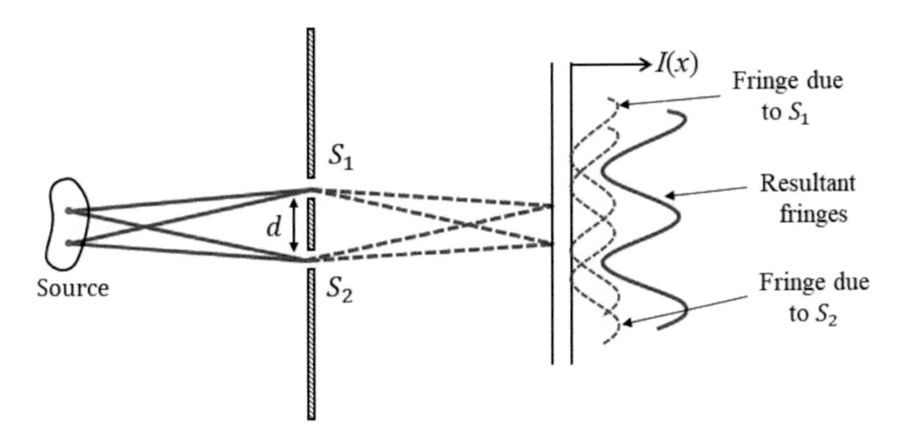

Figure A.2 Spatial coherence observation at the screen when the separation between two slits d is small.

In Eqn A.2 C_1, C_2 are constants. then the intensity $I(O)$ can be written as,

$$I(O) = |C_1|^2 < |U(s_1, t - \frac{r_1}{c})|^2 > + |C_2|^2 < |U(s_2, t - \frac{r_2}{c})|^2 >$$
$$+ C_1 C_2^* U(s_1, t - \frac{r_1}{c}) U^*(s_2, t - \frac{r_2}{c})$$
$$+ C_1^* C_2 U^*(s_1, t - \frac{r_1}{c}) U(s_2, t - \frac{r_2}{c}) \quad \text{(A.3)}$$

Substituting, $I_1(O) = |C_1|^2 < |U(s_1, t - \frac{r_1}{c})|^2 >$ and $I_1(O) = |C_2|^2 < |U(s_2, t - \frac{r_2}{c})|^2 >$ for the intensities produced separately by the slits 1, 2 respectively and $\Gamma_{12}(\tau) =< U(s_1, t + \tau) U^*(s_2, t) >$ representing cross correlation function due to interference effects we can simplify Eqn.A.3 to find the intensity on the screen as,

$$I(O) = I_1(O) + I_2(O) + C_1 C_2^* \Gamma_{12}(\frac{r_2 - r_1}{c}) + C_1^* C_2 \Gamma_{21}(\frac{r_1 - r_2}{c}) \quad \text{(A.4)}$$

Further, $\Gamma_{21}(-\tau) = \Gamma_{12}^*$ and $C_1 C_2^* = C_1^* C_2 = C_1 C_2$ and then Eqn A.4 can be simplified as,

$$I(O) = I_1(O) + I_2(O) + 2 C_1 C_2 Re \Gamma_{12}(\frac{r_2 - r_1}{c}) \quad \text{(A.5)}$$

Introducing self coherence functions Γ_{11} and Γ_{22} which are nothing but intensities of light beams entering at slits S_1, S_2 respectively and mutual coherence function Γ_{12} representing cross correlation function then complex degree of spatial coherence function γ_{12} will be,

$$\gamma_{12} = \frac{\Gamma_{12}}{[\Gamma_{11}(0)\Gamma_{22}(0)]^{\frac{1}{2}}} \quad \text{(A.6)}$$

where the complex degree of coherence function γ_{12} varies from 0 to 1 as,

$$0 \leq |\gamma_{12}(\tau)| \leq 1 \quad \text{(A.7)}$$

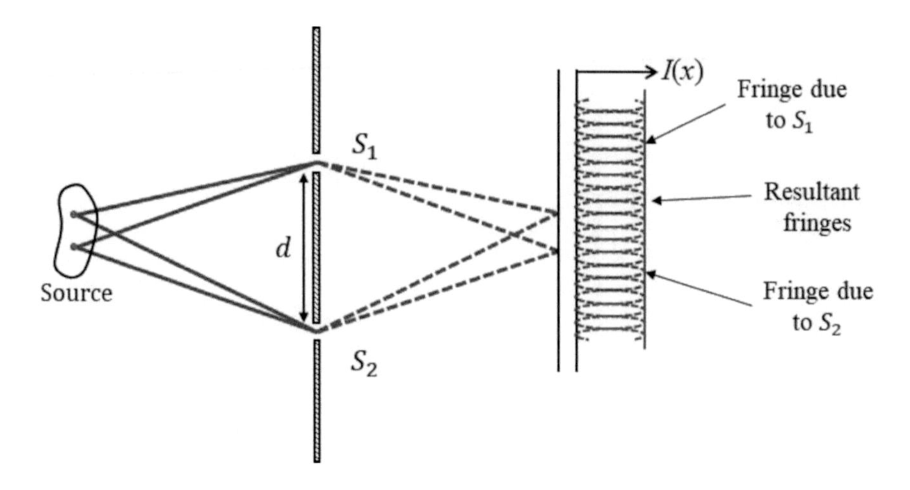

Figure A.3 Spatial coherence observation at the screen when the separation between two slits *d* is large.

The intensity shown in Eqn a.5 can be further simplified interms of intensities as,

$$I(O) = I_1(O) + I_2(O) + 2\sqrt{I_1(O)I_2(O)}Re\Gamma_{12}(\frac{r_2 - r_1}{c}) \tag{A.8}$$

with $I_1(O) = C_1^2\Gamma_{11}(0)$ and $I_2(O) = C_2^2\Gamma_{22}(0)$ respetively. Now, high visibility fringes on the screen can be observed for zero pathlength difference that is $(r_1 - r_2 = 0)$ and the interference fringes can have classical visibility ν equal to,

$$\nu = \frac{2\sqrt{I_1(O)I_2(O)}}{I_1(O) + I_2(O)}\gamma_{12}(0) \tag{A.9}$$

The value of $\gamma_{12}(0)$ represents the cross correlation co-efficient of waveforms from slits $U(s_1,t), U(s_2,t)$ respectively and when the distance between two slits d is changed, then $\gamma_{12}(0)$ varies from 0 to 1. This variation describes the spatial coherence of light beam incident on the slits.

A.3 TEMPORAL COHERENCE

A.3.1 INTRODUCTION

Consider $U(p,t)$ represents complex disturbance of light beam at a point in space p and at time t and $E(p,t)$ represents its complex amplitude(envelope) function of propagating light beam. As $U(p,t)$ has a finite bandwidth $\Delta\nu$ associated with it, one can expect that the amplitude and phase of $E(p,t)$ change when light beam propagates, at the rate determined by $\Delta\nu$. Suppose, if this complex field amplitude and phase does not change from time t to $(t+\tau)$, then the beam is said to have temporal coherence and $E(p,t)$ and $E(p,t+\tau)$ are said to be highly correlated. This temporal coherence phenomena can be explained using a Michelson Interferometer. Consider Fig. A.4 in which the geometry of a Michelson interferometer is shown. The light beam from the source is expanded and collimated. The collimated beam is split up in to two parts by a bemasplitter. The first part of beam travels directly to Mirror M_1 and gets reflected back and simultaneously the other part of beam falls on Mirror M_2 and gets reflected back. These two beams interfere at the beam splitter and it can be seen on the screen or detector as shown in Fig. A.5.

A.3.2 THEORETICAL EXPLANATION

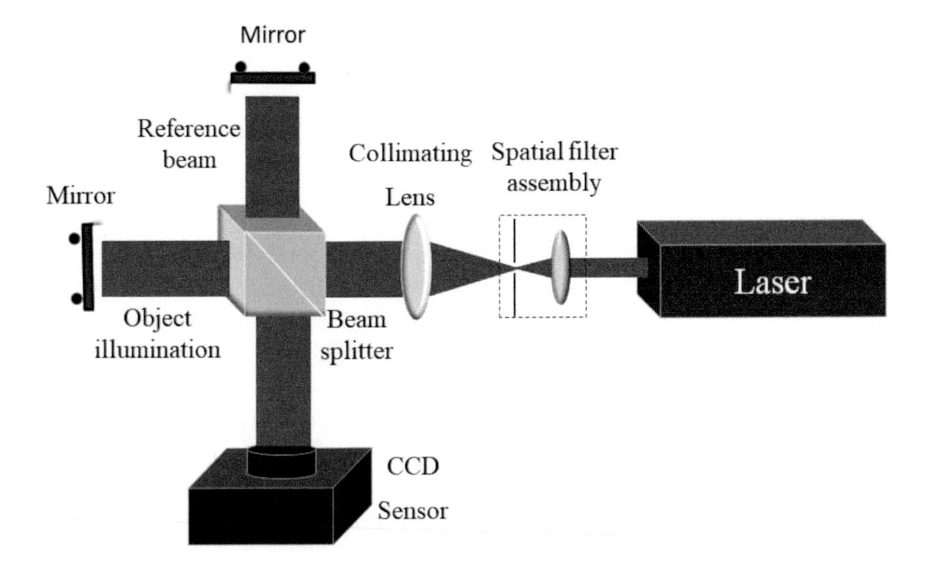

Figure A.4 Michelson interferometer geometry for temporal coherence measurement

Fig. A.4 shows a typical MIchelson Interferometer setup normaly students arrange in an optical laboratory. Considering that the source is not a laser then the beam reflected by mirror M_1 is $U(t,\frac{2d}{c})$ and by Mirror M_2 equal to $U(t)$ respectively. The distance d is relative path difference required for the formation of interference fringes by the detector when these two beams interfere. Assuming $\frac{2d}{c}$ is the time delay suffered by the light reflected by mirror M_1 then the total intensity at the detector will be,

$$I_t = < |C_1U(t) + C_2U(t + \frac{2d}{c})|^2 > \tag{A.10}$$

where, C_1, C_2 are the losses incurred by two beams along two paths respectively.

Further expanding above equation we get,

$$I_t = C_1^2 < |U(t)|^2 > +C_2^2 < |U(t+\frac{2d}{c})|^2 + C_1C_2 < U(t)U^*(t+\frac{2d}{c}) >$$
$$+C_1C_2 < U(t)^*U(t+\frac{2d}{c}) > \quad \text{(A.11)}$$

Substituting $I_0 =< |U(t)|^2 >=< |U(t+\frac{2d}{c}|^2 >$ and for cross correlation function $\gamma(\tau) =< U(t)U(t+\frac{2d}{c})^* >$, then Eqn A.12 becomes,

$$\begin{aligned} I_t &= (C_1^2+C_2^2)I_0 + C_1C_2\gamma(\frac{2d}{c}) + C_1C_2\gamma^*(\frac{2d}{c}) \\ &= (C_1^2+C_2^2)I_0 + 2C_1C_2Re[\gamma(\frac{2d}{c})] \end{aligned} \quad \text{(A.12)}$$

If $I_0 = \gamma(0)$ represents self coherence function and $\gamma(\tau)$ represents mutual coherence function of two time delayed beams then the complex degree of coherence function is given by,

$$\Gamma(\tau) = \frac{\gamma(\tau)}{\gamma(0)} \quad \text{(A.13)}$$

where, $\Gamma(\tau)$ is known as complex degree of coherence light. With these substitutions the total intensity recordedat the detector of Michelson Interferometer becomes,

$$I_t = (C_1^2+C_2^2)I_0[1+\frac{2C_1C_2}{(C_1^2+C_2)^2}Re\Gamma(\frac{2d}{c})] \quad \text{(A.14)}$$

The complex degree of coherence function can be expressed in following manner by introducing center frequency term ν and if the losses along two paths are same i.e $C_1 = C_2 = C$ then Eqn A.14 can be given by,

$$I_t = 2C^2I_0[1+\Gamma(\frac{2d}{c})cos[2\pi(\frac{2d}{c}) - a(\frac{2d}{c})] \quad \text{(A.15)}$$

If the realtive path difference between two beams is $d = 0$ then $\Gamma(\frac{2d}{c}) = 1$ and $a(\frac{2d}{c}) = 0$. Then the intensity at the detector will vary from $4C^2I_0$ to $2C^2I_0$. As the distnace d increases from 0 then the visibility of interference fringes falls as the interference fringes suffer a phase modulation equal to $a(\frac{2d}{c})$ which is directly related to coherence length of light source. The visbility of interference fringes can be given by,

$$V = \frac{(I_{max} - I_{min})}{(I_{max} + I_{min})} \quad \text{(A.16)}$$

In Eqn.A.16 I_{max} represents maximum intensity and I_{min} represents minimum intensity. For zero path difference between two beams we get $I_{max} = 1$ and for incoherent addition the intensity $I_{min} = 0$ which gives visbility a maximum value of 1. Eqn. A.16 clearly shows that as path difference grows, the visibility falls to zero which determines clearly the limit of temporal coherence. Finally Fig. A.5 shows the Michelson inteferometer fringes for a collimated He-Ne laser beam of wavelength 6328 A which has very large coherence length of few meters.

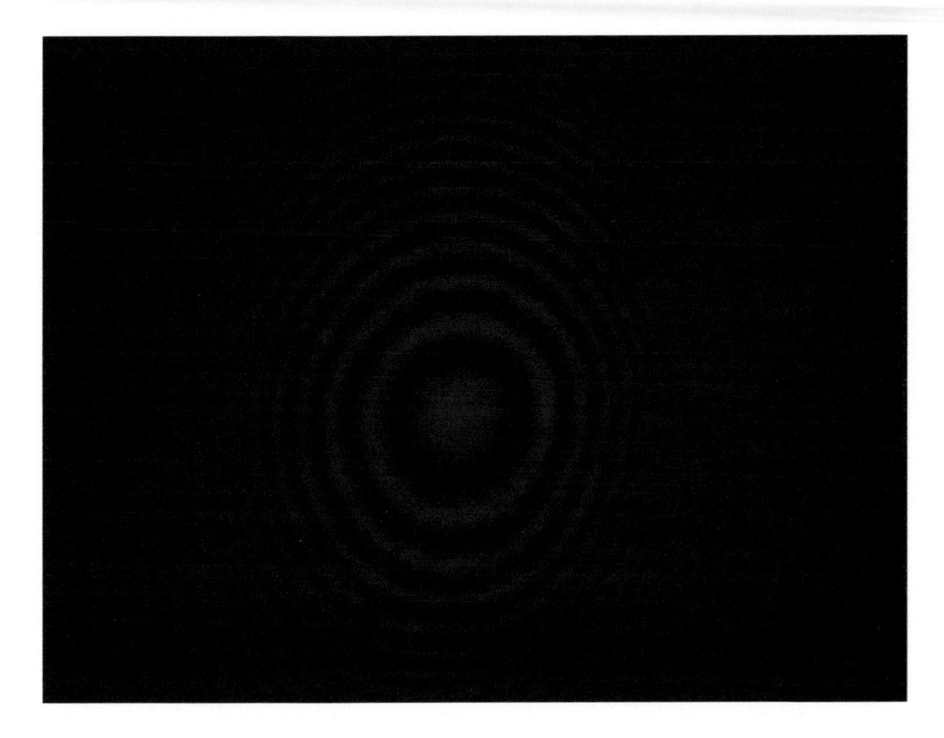

Figure A.5 Circular interference fringes of a Michelson interferometer.

B Rainbow holography

Normally, a conventional transmission hologram of a 3D object which is recorded using a particular wavelength λ_1 can be reconstructed and viewed using another light of wavelength λ_2. But, as we described in the section holographic magnifications(section 1.1.3) the position and magnifications varies with the wavelength of reconstructed beam. In this case therefore white light can not be used for reconstruction except for certain type of recording medium like photopolymers which is a recent development in holographic recording medium. In order to make holographic stickers or to use in advertisements for viewing in white light a technique called rainbow holography is used. Rainbow holograms are recorded using monochromatic light on a specially prepared holographic plate and are reconstructed or viewed using white light. Fig. B.1 shows a typical experimental set-up used to record a rainbow

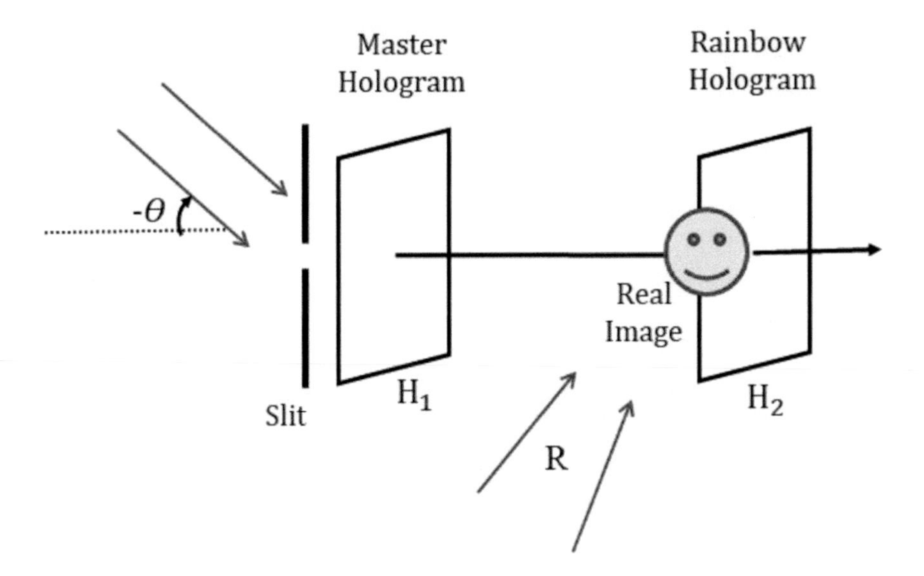

Figure B.1 Typical experimental geometry for recording master hologram for rainbow holography.

hologram. In that master hologram is recorded using conventional transmission hologram method. In Fig. B.1, H_1 is a master hologram of an object and is illuminated by the phase conjugate beam of its reference beam. In that geometry, unlike conventional hologram while reconstruction, an opaque screen with a narrow horizontal slit is placed infront of master hologram(shown in Fig. B.1). The opaque screen with slit is required to reduce the image resolution and brightness. The real image of the object on reconstruction is shown in Fig. B.1. A holographic plate(H_2) to record rainbow hologram is kept close, just behind the real image of master hologram. The hologram of this real image of master hologram is recorded using a convergent reference beam which is as shown in Fig. B.1. Care must be taken to use same light beam which was used to record the master hologram H_1. Now, the recorded holographic plate H_2 after wet processing gives the rainbow hologram of original object. This

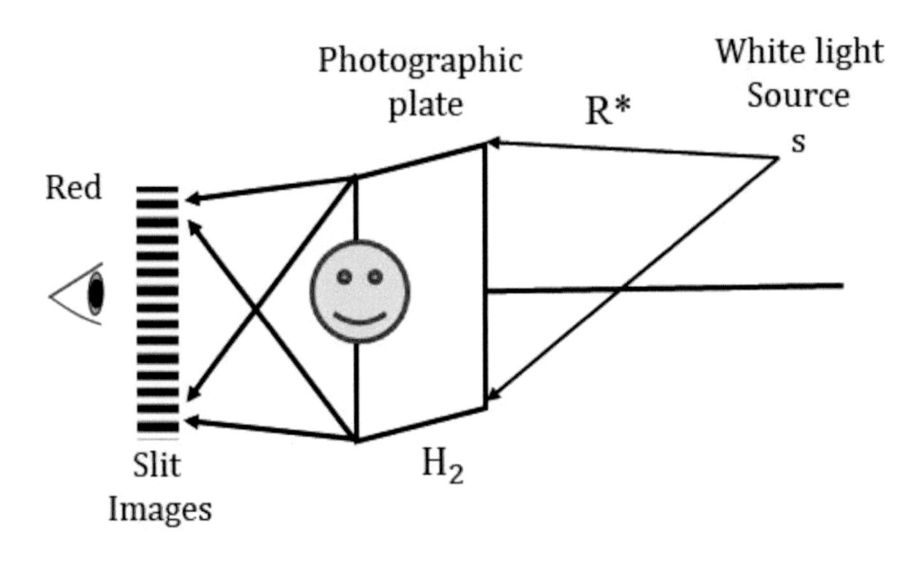

Figure B.2 Reconstruction of recorded rainbow hologram using a white light.

can be reconstructed using a divergent white light and it should propagate opposite to the convergent reference beam as shown in Fig. B.2. This rainbow hologram can generate multiple images in all colours of white light that is from red to violet. These colored images lie very close to rainbow hologram and they are real orthoscopic images of original object. Also, they are real images of pseudoscopic image formed by the master hologram. Further the rainbow hologram generates real images of slits in all colors of white light but certain distance away from it. This rainbow holography principle is used in present day holographic advertisements and stickers.

C Anisotropic self-diffraction

In case of the anisotropic self-diffraction process the polarization of diffracted beam will be rotated by some angle with respect to the transmitted object beam polarization. This happens in photorefractive crystals like BSO/BTO when the input face of the crystal is [110] plane and hologram vector coincides with [110] axis[30] and is shown in Fig 4.3. In this geometry when both object and reference beams interfere in the photorefractive $Bi_{12}SiO_{20}$(Bismuth Silicon Oxide), the amplitude of the phase grating produced inside the crystal is a maximum for light linearly polarized along and perpendicular to an axis bisecting the [001] and [110] crystal axes (i.e., at 45^0 to the grating vector K_g). This occurs due to the electrically induced birefringence of the crystal under the influence of an internal diffusion field. The shifted phase gratings produced by these polarizations are of opposite signs to that for writing- beam polarizations(Object and reference) perpendicular to the plane of inci- dence. If we ignore the optical activity, the crystal acts on the read out beam as a half-wave plate[30]. This results in linear polarization of the self-diffracted beam from such a phase grating rotated with respect to the transmitted beam. In considering the case in which there is optical activity, a maximum polarization separation of 90_0 between self-diffracted and transmitted polarizations is obtained when the reference-readout beam polarization is parallel to the [001] axis at the center of the crystal. Thus the transmitted beam can be canceled completely, which isolates the self-dif- fracted beam and thereby increases the image contrast of hologram to a greater extent. The angle of rotation of polarization plane of diffracted beam with respect to readout beam depends upon orientation of readout beam. If the plane of polarization of diffracted beam is rotated 90^0 with respect to readout beam then the plane of polarization of readout beam should be normal to centre of [110] crystallographic axis. This rotation of diffracted beam with respect to transmitted beam in an an-isotropic self diffraction can be utilised to separate the direct beam from diffracted object beam by putting an analyzer after the crystal. The physical process of isotropic and anisotropic self diffration phenomena is shown in Fig. C.1

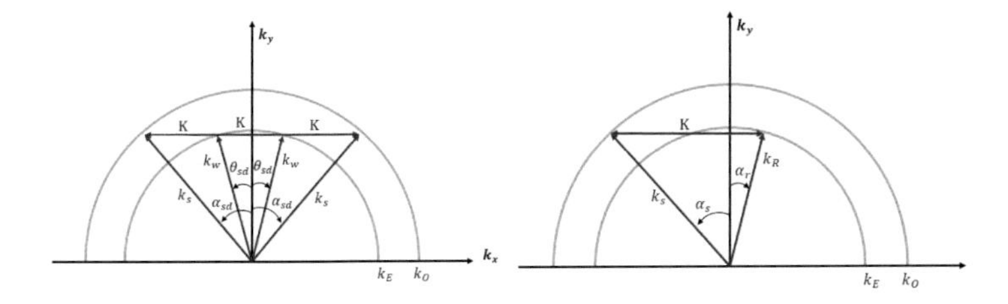

Figure C.1 a) Isotropic diffraction in uni-axial crystal b) An-isotropic self-diffraction in photorefractive crystal.

a, b respectively. If the transmitted beam is polaized linearly along x-direction then the diffracted beam from photorefractive crystal will be along y-direction. Now, by rotating the analyzer one can easily suppress the directly transmitted beam through the crystal and due to that the dynamic hologram obtained via diffracted beam can be separated for high efficiency. Thus simultaneously efficient readout and recording of hologram can be obtained.

D Van Cittert–Zernike theorem

D.1 INTRODUCTION

The Van Cittert–Zernike theorem describes the character of propagation of mutual intensity function from an incoherent source and thus becomes a very important tool in modern optical imaging. This theorem becomes very important for almost all optical incoherent sources for imaging. This theorem was first demonstrated by Van Cittert and Zernike[47]. This theorem also was used by Takeda et.al[46] to demonstrate coherence holography described in chapter 6 of this book.

D.2 THEORETICAL EXPLANATION

In case of quasimonochromatic light[47], the propagation of mutual intensity is given by,

$$\mathbf{V}(P_1,P_2) = \int\int_{\Sigma}\int\int_{\Sigma}\mathbf{V}(q_1,q_2)e^{[-i\frac{2\pi}{\lambda}(r_2-r_1)]}\frac{\Gamma(\theta_1)}{\lambda r_1}\frac{\Gamma(\theta_2)}{\lambda r_2}dA_1dA_2 \tag{D.1}$$

The term $\mathbf{V}(q_1,q_2)$ represents initial state of coherence of the quasimonochromatic beam and introducing special case for an incoherent source as,

$$\mathbf{V}(q_1,q_2) = \kappa I(q_1)\delta(|q_1-q_2|) \tag{D.2}$$

the Eqn.D.1 becomes,

$$\mathbf{V}(P_1,P_2) = \frac{\kappa}{\lambda^2}\int\int_{\Sigma}I(q_1)e^{[-i\frac{2\pi}{\lambda}(r_2-r_1)]}\frac{\Gamma(\theta_1)}{\lambda r_1}\frac{\Gamma(\theta_2)}{\lambda r_2}dA \tag{D.3}$$

Fig. D.1 shows the required geometry for this derivation of Van Cittert–Zernike theorem. For further simplifying Eqn.D.3 we assume that i) The areas of source and observation region shown in Fig. D.1 is much smaller than the separating distance z between them ($\frac{1}{r_1}.\frac{1}{r_2} = \frac{1}{z^2}$), ii) Angles are small as $\Gamma(\theta_1) = \Gamma(\theta_2)$ and with these assumptions, the mutual intensity from the quasimonochromatic incoherent source observed at the observation region as shown in Fig. D.1 is given by,

$$\mathbf{V}(P_1,P_2) = \frac{\kappa}{(\lambda z)^2}\int\int_{\Sigma}I(q_1)e^{[-i\frac{2\pi}{\lambda}(r_2-r_1)]}dA \tag{D.4}$$

Considering Fig. D.1 where, the source and observation planes are separated by a distance z and using the paraxial approximation we get,

$$\begin{aligned} r_1 &= \sqrt{z^2+(x_1-\xi)^2+(y_1-\eta)^2} = z+\frac{(x_1-\xi)^2+(y_1-\eta)^2}{2z} \\ r_2 &= \sqrt{z^2+(x_2-\xi)^2+(y_2-\eta)^2} = z+\frac{(x_2-\xi)^2+(y_2-\eta)^2}{2z} \end{aligned} \tag{D.5}$$

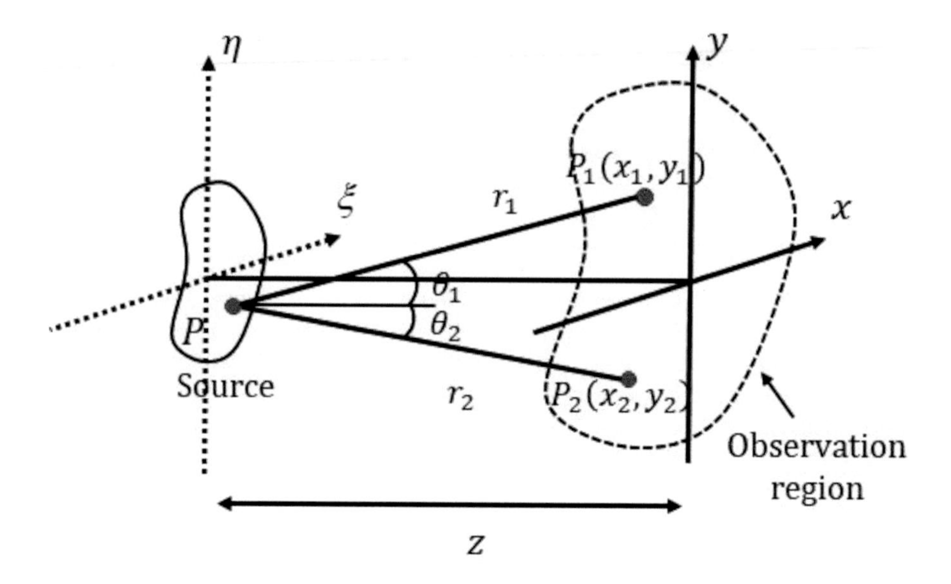

Figure D.1 Geometry for demonstration of Van Cittert–Zernike theorem.

substituting $\Delta x = (x_2 - x_1), \Delta y = (y_2 - y_1)$ and the intensity value of source $I(\xi, \eta) = 0$ outside the area of source i.e if the co-ordinates (ξ, η) lies outside the source surface region the final form of Van Cittert–Zernike theorem becomes,

$$\mathbf{V}(\mathbf{x_1}, \mathbf{y_1}, \mathbf{x_2}, \mathbf{y_2}) = \frac{\kappa e^{-i\psi}}{(\lambda z)^2} \int \int_{-\infty}^{+\infty} I(\xi, \eta) e^{\left[\frac{-i2\pi}{\lambda z}(\Delta x \xi + \Delta y \eta)\right]} d\xi d\eta \tag{D.6}$$

In above expression the phase factor ψ is given by,

$$\psi = \frac{\pi}{\lambda z}[(x_2^2 + y_2^2) - (x_1^2 + y_1^2)] = \frac{\pi}{\lambda z}(\rho_2^2 - \rho_1^2) \tag{D.7}$$

where ρ_1, ρ_2 represents the distances of points (x_1, y_1) and (x_2, y_2) from origin $(0,0)$ on the optical axis at the observation plane (x, y) as shown in Fig. D.1 respectively. Now, the complex coherent factor is given by,

$$\mu(x_1, y_1; x_2, y_2) = \frac{e^{-i\psi} \int \int_{-\infty}^{+\infty} I(\xi, \eta) e^{\left[\frac{-i2\pi}{\lambda z}(\Delta x \xi + \Delta y \eta)\right]} d\xi d\eta}{\int \int_{-\infty}^{+\infty} I(\xi, \eta) d\xi d\eta} \tag{D.8}$$

It may be noted here that, in most of practical applications involving incoherent sources we can sfaely put $I(x_1, y_1) = I(x_2, y_2)$ and the term $\mu(x_1, y_1; x_2, y_2)$ represents visibility of Young's double slit experiment.

D.3 INTERPRETATION OF VAN CITTERT–ZERNIKE THEOREM

The Van Cittert–Zernike theorem shown mathematically by Eqn. D.6 states that the mutual intensity $V(x_1, y_1; x_2, y_2)$ can be obtained by two dimensional Fourier Transformation of the intensity distribution $I(\xi, \eta)$ across the source itself. This relationship can be compared with the relationship observed between the light field across a coherently illuminated aperture and the field observed at the Fraunhoffer diffraction pattern of the aperture though the physical quantities involved are different. In this the intensity distribution $I(\xi, \eta)$ is analogous to the field across the aperture and the mutual intensity function $V(x_1, y_1; x_2, y_2)$ is analogous to the field at the Fraunhoffer diffraction pattern and this analogy is only mathematical one and are different for physical situation. Also, the Fourier Transform relationship between $V(x_1, y_1; x_2, y_2)$ and $I(\xi, \eta)$ holds over a wide range of distances. The experimental use of Van-Zittert Zernike theorem[47] is well established in Coherence holographic experiments developed by Takeda et.al[46].

References

1. D. Gabor, “A new microscopic principle”, *Nature,* **161,** 777-778 (1948).
2. D. Gabor, “Microscopy by reconstructed wavefront”, *Proc. Royal Society A,* **197,** 454-487 (1949).
3. E. Leith and J. Upatnieks, “Reconstructed wavefronts and communication theory”, *JOSA,* **52**, 1123-1128 (1962).
4. Y N Denisyuk, “Photographic reconstruction of the optical properties of an object in its own scattered field”, *Sov. Phys. Dokl,* **7:543**, (1962).
5. E. N. Leith and J. Upatnieks, “Wavefront Reconstruction with Continuous-Tone Objects,” *JOSA,* **53**, 1377 (1963).
6. Joseph W Goodman “Introduction to Fourier Optics”, *McGraw Hill,* New York (1968)
7. Robert J Collier, Christoph B Burckhardt, Lawrence H.Lin, “Optical Holography, “*Academic Press*, New York (1971).
8. H. J Caulfield (Ed), “Hand book of Optical Holography, “*Academic Press*, New York, (1979).
9. Francis T. S. Yu, “Optical Information Processing”, *John wiley & sons*, New York, (1983).
10. P. Hariharan, “Optical holography”, *Cambridge University Press*, Cambridge, (1984).
11. K K Sharma “Optics”, *Academic Press,* New Delhi (2006).
12. G. Sirat and D. Psaltis, “Conoscopic holography,” *Opt. Lett.* **10,** 4-6 (1985).
13. G. Sirat and D. Psaltis, “Monochromatic incoherent light holography,” *U.S. patent* 4, 602, 844 (July 29, 1986).
14. G. Y. Sirat and D. Psaltis, “Conoscopic holograms,” *Opt. Commn.***65,** 243-249 (1988).
15. Gabriel Y. Sirat “Consocopic holography. I Basic principles and physical basis, “*Jl. Opt. Soc. Am. A***9,** No.1 70-83 (1992)
16. Gabriel Y. Sirat “Conoscopic holography. II Rigorous derivation, “*Jl. Opt. Soc. Am. A***9,** No.1 84-90 (1992)
17. K. Buse and M Luennmann “# D Imaging : Wavefront sensing utilizing a birefringent crystal, “*Physical Review Letters* **85,** No.6 3385-87 (2000)
18. A. W. Lohmana and D. P. Paris, “Binary Fraunhoffer holograms generated by computer”, *Appl. Optics* **6** 1739-48 (1967)
19. W. H. Lee, “Sampled Fourier Transform hologram generated by computer”, *Appl. Optics* **9** 639-43 (1970)
20. W. H. Lee, “Binary synthetic hologram”, *Appl.Optics* **13** 1677-1682 (1974)
21. L.P Yaroslavsky, “Digital Holography and Digital Image Processing”, *Kluwer Academic Publishers*, 2004
22. A.Vijayakumar, B.J Jackin and P.K Palanisamy “Computer generated Fourier holograms for undergraduate optics laboratory” *Physics Education* **28** 4 (Oct-Dec 2012)
23. M. Gower, “Dynamic holograms from crystals”, *Nature,* **316,** 12 (1985).
24. T. J. Hall, R Jaura, L. M Connors and P.D Foote, “ The photorefractive effect - A Review *Prog. in Quantum Electronics*, **10**, 77 (1985)
25. S.I. Stepanov and M.P Petrov, “Photorefractive materials and applications”, *Topics in Appl. Physics: P Gunter and J.P. Huignard (Eds),* Springer (1988).

26. P.Yeh, "Introduction to photorefractive non-linear optics", John Wiley & Sons, New York(1993)
27. F.S. Chen, J.T. La Macchia and D.B Fraser, "Holographic storage in Lithium Niobate(LiNBO3)," *Appl. Phys. Lett*, **13** 223 (1968)
28. S.I. Stepanov and M.P Petrov, "Adaptive holographic interfereometers operating through self-diffraction of recording beams in photorefractive crystals", *Opt. Commun*, **53** 292 (1985)
29. A.A. Kamshalin, E.V Mokrushina and M.P Petrov, "Efficient unstationary holographic recording in photorefractive crystals under external alternating electric fields", *Opt. Engg*, **28** 580-88 (1989)
30. R.C. Troth and J.C. Dainty "Holographic interferometer using an-isotropic self-diffraction in $Bi_{12}SiO_{20}$ *Optics Letters* **16** 53-55, 1991
31. T. S Huang, "Digital holography", *Proc. IEEE*, **59(9)** 1335-46(1971)
32. M.A. Kronrod, N.S. Merzlyakov and L.P. Yaroslavskii, " Reconstruction of hologram with a computer", *Sov. Phys-Tech. Phys* **17** 333-334, (1972)
33. L.P. Yaraoslavsky and N.S. Merzlyakov, "Methods of digital holography", *Consultants bureau* New York (1980)
34. U. Scanners and W. Juptner "Direct recording of holograms by a CCD target and numerical reconstruction", *Appl. Opt.,* **33(2)**, 179-81 (1994)
35. U. Scanners and W. Juptner "Direct recording and reconstruction of holograms in hologram interferometry and shearography", *Appl. Opt.,* **33(20)**, 4373-4377 (1994)
36. Thomas Kries, "Handbook of Holographic Interferometry", **Wiley-VCH**, 2005 Weinheim
37. Thomas M Kries and Werner P.O. Juptner "Suppression of the DC term in Digital holography", **Opt. Engg (SPIE)**, **36(8)** 2357-60 (1997)
38. I. Yamaguchi and T. Zhang "Phase shifting digital holography", *Opt. Lett.,* **22(16)**, 1268-70 (1997)
39. T. H. Chyba, L. J. Wang, L. Mandel, and R. Simon "Measurement of the Pancharatnam phase for a light beam", *Opt. Lett.,* **13(7)**, 562-64 (1988)
40. I. Yamaguchi, T. Matsumura and J.I. Kato "Phase shifting color digital holography", *Opt. Lett.,* **27(13)**, 1108-10 (2002)
41. D. Gabor and W. P. Goss "Interference Microscope with Total Wavefront Reconstruction," *Jl. Opt. Soc. Am.,* **56(7)**, 849-858 (1966)
42. Jung-Ping Liu and Ting-Chung Poon "Two-step-only quadrature phase-shifting digital holography", *Opt. Lett.,* **34(03)**, 250-252 (2009)
43. P. Hariharan and Maitreyee Roy "A geometric phase interferometer", *Journal Of Modern Optics.,* **39(9)**, 1811-15 (1992)
44. Jun-ichi Kato and Ichirou Yamaguchi "Multicolor digital holography with an achromatic phase shifter", *Opt. Lett.,* **27(16)**, 1403-05 (2002)
45. Boaz Jessie Jackin, C.S Narayanamurthy and Toyahiko Yatagai "Geometric phase shifting digital holography", *Optics. Lett.,* **41(11)** 2648-51 (2016)
46. Mitsuo Takeda, Wei Wang, Zhihui Duan, and Yoko Miyamoto", *Opt. Exp.,* **13(23)**, 9629-35 (2005)
47. J .W. Goodman *Statistical Optics*, **Chapter 5**, John Wiley & Sons, New York, 1985

Index

Note: Locators in *italics* represent figures and **bold** indicate tables in the text.